D0847457

Modal Thinking

by the same author

ATTENTION

G. E. MOORE

Modal Thinking

ALAN R. WHITE

Ferens Professor of Philosophy
in the University of Hull

CORNELL UNIVERSITY PRESS

Ithaca, New York

For information address Cornell University Press, 124 Roberts Place, Ithaca, New York 14850.

First published 1975 by Cornell University Press.

International Standard Book Number 0–8014–0991–8
Library of Congress Catalog Card Number 75–16811

Printed in Great Britain

Contents

Preface

Much attention has recently been given to formal modal logic and to various of its problems, but little to the actual nature of those concepts of our everyday thinking for which this formal logic was ostensibly designed. This book is a preliminary attempt to investigate these concepts.

I take this opportunity of expressing my general gratitude to all those friends and colleagues, at Hull and elsewhere, too numerous to mention by name, whose criticisms on particular chapters have enabled me to remove many earlier blemishes. A special debt is due to those who undertook the laborious task of commenting on the entire work in an earlier draft. For this I gladly thank Jonathan Cohen, Peter Geach, Roger Montague, Gilbert Ryle, Aaron Sloman and my students and colleagues at Temple University.

Some of the material on which this book is based previously appeared in *Analysis*, the *American Philosophical Quarterly*, the *Proceedings of the Aristotelian Society*, the *British Journal of Educational Studies* and the *Royal Institute of Philosophy Lectures* and I should like to thank the editors and the publishers concerned for their permission to reuse it here.

Introduction

Among the quite fundamental notions which we use in our thinking are those sometimes referred to as *modals.* Their range is a matter of dispute. Grammarians, for instance, use several criteria[1]* for designating modal verbs as a sub-class of auxiliary verbs. Thus, they sometimes call modals all auxiliary verbs in English except 'be', 'have' and 'do'. Their list of modals, therefore, comprises 'will', 'shall', 'dare', 'used', 'can', 'may', 'must', 'have to', 'ought', and 'need'. Sometimes 'need' and 'dare' are classed as catenative rather than modal and 'used' is regarded as only a quasi-auxiliary. Various adjectives, adverbs, etc., are sometimes[2] called 'modals' on semantic grounds. Logicians,[3] on the other hand, usually restrict modality to the notions of *possibility* and *necessity* and, perhaps, any notion definable in terms of these. Lately, however, they have added 'may', 'must' and 'ought' as deontic modals and have tended to widen modality to include epistemic and temporal ideas. Indeed, a very recent move[4] has widened the category to include all these and also so-called 'boulomaic' modals (e.g. it is hoped or feared that), 'causal' modals (e.g. it is probable that), etc., on the ground that it was a practice of Aristotle's medieval commentators to use the idea of a mode to cover any adverbial qualification of a verb. One medieval logician, William of Shyreswood, characterized a modal statement as one which 'says how the predicate inheres in the subject'. But any appeal to the etymology of 'modal' from 'modus' allows far too hospitable a range of modals.

For my purpose, I shall select as the basic modal notions those expressed by the modal auxiliaries 'can', 'may', 'must', 'ought', and 'need' together with some other notions which we shall see are closely related to them. For instance, related to 'can' are the

* References to this chapter begin on p. 3.

ideas of *possibility*, *ability* and *power*; related to 'may' are those of *possibility*, *probability* and *certainty;* related to 'must' and 'need' is the idea of *necessity* and related to 'must' and, allegedly, to 'ought' is that of *obligation.*

The modal notions are important both in themselves and because misunderstanding of them lies at the root of several of the oldest philosophical problems. Thus, scepticism is based on the feeling that nothing can be known unless the possibility of its being otherwise has been ruled out; while determinism usually enshrines the belief that it is not possible for anything to be otherwise than it is. The problem of free will is the problem whether anyone could have done something other than what he did do; a problem which many recent philosophers have assumed or argued to be the problem whether he would have done otherwise, if he had wanted or chosen to. Certainty is often confused with necessity because of failure to notice that the possibility expressed by 'may' differs from that expressed by 'can'. Many theories of probability are based on the subjective interpretation of the idea of possibility. The nature of morality is misunderstood because the notion expressed by 'ought' is assimilated to the notion of obligation. The confusion of needs with wants is a root cause of fallacies in psychology, politics and education. Responsibility is thought to be based on the presupposition that someone can necessarily do what he ought to do. Of many queer philosophical views it is true to say that there is modality in their madness.

Controversy about the nature of the modals themselves has usually centred on the question whether they are subjective, that is, used to express something about the user, or objective, that is, used to state something about the features of that to which they refer. Many grammarians, for instance, connect modality with mood and regard it—like moods of the verb—as a way of marking the 'attitude of the speaker';[5] while others interpret it objectively as asserting 'a specific relation between that event and the factual world'.[6] 'Modal subjectivism' could be used rather loosely to cover a wide and somewhat heterogeneous group of philosophical views, such as: Hume's contention that 'necessity is something that exists in the mind, not in objects';[7] its modern corollary which analyses *ought* and *must* in terms of an agent's reasons;[8] prescriptivist theories of *ought* and *obligation* which tend to deny the applicability of truth and falsity to moral pronouncements;[9] the recent contention that 'the term "need" is mainly normative';[10] subjectivist[11] and attitudinal[12] views about probability as well as much talk about epistemic possibility and certainty.[13] By contrast, objectivist analyses of modal concepts—of which Leibniz and Mill seem to have been advocates—have been out of favour recently, perhaps because of the feeling that they are committed to the assumption that modal words are names of mysterious features of the world such as necessity, probability and obligation.

It will be best to leave the subjectivist's controversy with the objectivist unresolved until we have examined the actual use of the modals, more particularly since such an examination has rarely been conducted with much thoroughness either in philosophy[14] or in grammar.[15] We shall, indeed, uncover many false assumptions about their use.

My aim, therefore, will be first, to get clear about the use of the modal auxiliaries and related concepts; secondly, to show, at the appropriate stages, how such clarity provides a clue to the solution of various traditional philosophical problems; and, finally, to characterize the general nature of modal thinking. It should be borne in mind that because the modals are among the most fundamental of all our concepts, their analysis can consist, as some grammarians and philosophers have stressed,[16] only in showing their relations to other concepts and not in revealing some more basic concepts as their constituents.

NOTES

1. E.g. Palmer's (ch. 2) criteria for auxiliaries are a paired negative form (e.g. mightn't), inversion with subject (e.g. Can he come?), repetition (e.g. I ought to and so ought you/I ate it and so did you), emphatic affirmation (e.g. He *must* do it). In addition, the modals take the infinitive of the verb. Twaddell's, p. 2, criteria for modals, in addition to those for auxiliaries, are absence of 's', absence of full 'past' syntax, and spatial precedence over the auxiliaries.
2. E.g. Poutsma, quoted in Greenbaum, p. 100.
3. E.g. Hughes and Cresswell, p. ix; Carnap (1956), ch. V; Kneale (1962), *passim*.
4. Rescher, ch. 4; cp. von Wright (1951b); Prior (1962), ch. III.i 5.

5. E.g. Lyons, pp. 307–13; Greenbaum, chs. 5 and 8, includes such adverbs as 'possibly', 'probably', 'certainly' under 'attitudinal disjuncts' and Poutsma, quoted by Greenbaum, says of what he calls 'modal adverbs' that they express 'the speaker's attitude'.
6. E.g. Joos, ch. 6; cp. Palmer, chs 6 and 7 and Lebrun. Ehrman, ch. 9, combines both subjective and objective.
7. *Treatise*, I.iii 14.
8. E.g. D. G. Brown, pp. 5, 70–80, 102; Edgley, ch. 4. 10.
9. E.g. Hare (1952).
10. E.g. Peters (1958), pp. 17–19; Benn and Peters, pp. 141 ff.; Hirst and Peters, pp. 32 ff.
11. E.g. Carnap (1944–5).
12. E.g. Toulmin; Fogelin.
13. E.g. Moore (1959), ch. 10; (1962), pp. 184–8; Hacking (1967); Harré; Heidelberger.
14. Some preliminary work has recently been done, e.g., by Moore (1962); Harré; Hacking (1967); Ayers (1968) and R. Wertheimer.
15. Recent grammatical work, such as that of Ehrman and Lebrun, is rather naive on the meanings and functions of modal words.
16. Joos, ch. 6; Kneale (1962a), p. 125; Toulmin, ch. 1.

Chapter One

Possibility

The notion of *possibility* is used in two quite distinct constructions, one of which is commonly expressed in English by 'possible to' and the other by 'possible that' with the indicative. We can say 'It is possible to get from Hull to London in 3 hours', 'It is possible for milk to turn sour overnight', 'It is possible for a healthy man to lift nearly a hundredweight'; and we can also say 'It is possible that the window was broken yesterday', 'It is possible that he is in the garden', 'It is possible that it will rain before evening'.[1]*

The use of 'possible that' with the subjunctive—either in the plain form or the 'should form'—e.g. 'It is possible that a triangle should have three equal sides', is a variation on 'possible to' and is not co-ordinate with 'possible that' with the indicative. This is clear from the facts that (1) the use of 'that' with the subjunctive, as equivalent to an infinitive, is common with other words. Contrast, for example, 'It is best (right, only fair) that he (should) leave'—which is equivalent to 'It is best (right, only fair) for him to leave'—with 'It is best (right, only fair) that he is leaving'; (2) 'It is necessary that' with the subjunctive, e.g. 'It is necessary that a triangle (should) have three sides', is a variation on 'It is necessary to', e.g. 'It is necessary for a triangle to have three sides'; and (3) 'It is possible that' in the subjunctive-governing use is linked to 'It is necessary that' and not—as we shall see is the case with 'It is possible that' with the indicative—to 'It is certain that' or 'It is probable that'. We do not say 'It is certain (probable) that X (should) be so' just as we do not say 'It is certain (probable) for X to be so', but only 'It is certain (probable) that X is (was, will, would be) so'.

To say that it is possible for X to V—where 'V' is a variable

* References to this chapter begin on p. 17.

for any verb—is to stress the actual existence of a possibility, that is, what possibility actually exists; whereas to say that it is possible that X Vs is to stress the possible existence of an actuality, that is, what actuality possibly exists. Hence, I shall sometimes, for convenience, call the former 'existential' possibility and the latter 'problematic' possibility. In existential possibility nothing prevents X from Ving, while in problematic possibility something allows it. The former possibility indicates what is capable of being so, the latter what is likely to be so.

The difference between the actuality of a possibility and the possibility of an actuality accounts for differences in the ways in which expressions of the two possibilities are tensed. In the former, it is usually the possibility that is temporally qualified, in the latter it is that which is possible. Because problematic possibility is a present possibility that something past was or would have been so, that something present is or would be so or that something future will or would be so, we say 'It *is* possible that X did V, is Ving, will V or would have Ved, would be Ving or would V'. On the other hand, because existential possibility is a present, past or future actuality of a possibility, we say 'It *is*, *was*, *will be*, *would be* or *would have been* possible for X to V' or 'It is possible for X to have Ved or to be about to V'. We can say that it once was, but no longer is, possible for X to V, but not that it once was, but no longer is, possible that X did V.[2] It can only have once seemed possible that X did V. To say 'It *was* at that time possible that X *would* V' is not an exception, since 'was' here predates 'would'. Another way to put this difference is this. It *is*, *was* or *will be* possible for something to occur, e.g. for A to do something or for X to happen; whereas it only *is* possible that it *is* the case that something occurred, occurs or will occur. It cannot *have been* or *be about to be* possible that it *is* the case that . . . 'It has not yet been and never will be possible for A to V' makes sense, but not 'It has not yet been and never will be possible that A does V'. Similarly, it can rarely, frequently, sometimes, always or never be possible for A to V, but not that A does V.

Both kinds of possibility are implied by actuality; if something is so, it must be possible for it to be so and possible that it is so.

On the other hand, neither possibility implies actuality;

whether it is possible for something to be so or possible that it is so, it need not actually be so. The fact that one can legitimately argue from the impossibility of something to its non-existence—symbolically, $\sim Mp \rightarrow \bar{p}$—provides not the slightest reason for supposing that one can argue from the possibility of something to its existence—symbolically $Mp \rightarrow p$. Hence, the occurrence of negative ontological conclusions about round squares from the logical impossibility of a round square does not lend any respectability[3] to a positive ontological conclusion about God from the logical possibility of his existence.

Equally, neither possibility implies non-actuality; to say either that it is possible for X to V or possible that X Vs does not imply that X does not V.

Philosophers, however, have disputed both these assertions about what possibility implies. As regards the second assertion, no one, of course, has been so foolish as to suggest that either 'It was possible for X to V' or 'It is possible that X did V' implies 'X did not V', since they would, by contraposition, give the absurd result that 'X did V' implies 'It was not possible for X to V' and 'It is not possible that X did V'. It has, however, been held that 'It would have been possible for X to V' and 'It is possible that X would have Ved' each implies either that X did not V or that the speaker supposes that X did not V. What little plausibility this view has stems partly from a general thesis that 'would have' implies 'did not' and partly from a misunderstanding of the implications of 'could have' and 'might have', the terms in which the view is usually held. These points will be discussed in the next two chapters.

The opposition to the first assertion, namely that possibility does not imply actuality, has usually taken the contrapositive form that 'X did not V' implies both that it was not possible for X to V and that it is not possible that X did V.[4] Since the former implication is sometimes expressed as the deterministic thesis 'Nothing but what did happen could have happened' and the latter as 'Nothing can be improbable but true',[5] we can most appropriately examine the various sources of this common fallacious inference in our later discussion of determinism and of probability. Here, however, its misunderstanding of the notion of possibility needs to be shown. For this limited purpose we can deal simultaneously both with 'It was not possible for X to V' and with the quite different 'It is not possible that X did V'.

If 'X did not V' implied that it was not possible for X to V or implied that it is not possible that X did V, then, by contraposition, 'It was possible for X to V' and 'It is possible that X did V' would each imply that X did V. Apart from its intuitive implausibility, this supposition, taken together with the generally agreed view that 'X did V' implies both that it was possible for X to V and that it is possible that X did V, would give the result that 'X did V' is equivalent both to 'It was possible for X to V' and to 'It is possible that X did V'. And, similarly, that 'X did not V' is equivalent both to 'It was possible for X not to V' and 'It is possible that X did not V'. And this is very queer. Furthermore, if these were equivalent, then since 'X did V' and 'X did not V' are incompatible, then their alleged equivalents 'It was possible for X to V' (or 'It is possible that X did V') and 'It was possible for X not to V' (or 'It is possible that X did not V') would be incompatible. But, apart from our intuitive feeling that these are not incompatible, the opposite supposition leads to an absurdity. For, if 'It was possible for X to V' (or 'It is possible that X did V') were incompatible with 'It was possible for X not to V' (or 'It is possible that X did not V'), then 'It was possible for X to V' (or 'It is possible that X did V') would imply 'It was not possible for X not to V' (or 'It is not possible that X did not V'). But, as I shall show later, 'It was not possible for X not to V' implies 'It was necessary for X to V' and 'It is not possible that X did not V' implies 'It is certain that X did V'. Therefore, 'It was possible for X to V' would imply 'It was necessary for X to V' and 'It is possible that X did V' would imply 'It is certain that X did V'. And both of these are absurd.

We see, therefore, that it is incorrect to suppose that 'X did not V' implies either that it was not possible for X to V or that it is not possible that X did V. This mistaken supposition has in fact been confused with the correct thesis that 'X did not V' implies both that it was possible for X not to V and that it is possible that X did not V.

To say that 'It is possible for X to V' and 'It is possible for X not to V' are compatible is not, of course, to say that 'It is possible for X to V and 'It is not possible for X to V' are compatible. Nor has anyone, as far as I know, ever supposed otherwise.[6] Equally, to say that 'It is possible that X does V' and 'It is possible that X does not V' are compatible is not to say that 'It is possible that X does V' and 'It is not possible that X does V' are compatible. G. E. Moore, however, alleged[7] that the latter two are compatible. But that 'X does V' and 'X does not V' are both possible does not show that it is both possible and not possible that X does V. If it is not possible that X does V, then it follows that it is possible that X does not V, but from 'It is possible that X does not V', it does not follow that it is not possible that X does V. Moreover, since, as we shall see later, 'It is possible that X does V' if and only if 'It is not certain that X does not V' and 'It is not possible that X does V' if and only if 'It is certain that X does not V', then 'It is certain that X does not V' and 'It is not certain that X does not V' are—contrary to what Moore also said—equally incompatible. The fact that 'It is not certain that X does V' and 'It is not certain that X does not V' are compatible does not show that 'It is certain that X does V' and 'It is not certain that X does V' are compatible. The supposition that neither 'X does V' nor 'X does not V' is certain is different from the supposition that 'X does V' is both certain and uncertain. Indeed, Moore obviously thought, in his insistence at his lecture that it was certain that he was standing up in front of an audience, that he was disagreeing with Russell and other sceptics who suggested that this and many others things, which are universally believed, are not certain.

Modal logicians[8] have, of course, always allowed that a contingent proposition and its negation are both possible, but they do not make it clear whether by saying that 'A proposition p is possible' they mean that it is possible that a proposition p should be true (or for a proposition p to be true) or possible that a proposition p is true. Nor is Aristotle, who is often appealed to here,[9] altogether clear on this point. Aristotle's use of the word δυνατόν in *De Interpretatione* 18b–23a suggests that he argued only for the truth of 'It is possible for it to be and it is possible for it not to be'. Aristotle's words at 19a, namely δυνατόν ἐστι διατμηθῆναι κἀι οὐ διατμηθήσεται, which are sometimes wrongly translated either as 'It may be cut and it will not be cut'[10] or as 'It may be cut and it may not be cut',[11] ought to be translated as 'It can be (that is, it is possible for it to be) cut even though it will not be cut'.[12]

Despite their similar relation to actuality, existential possibility and problematic possibility are very different. First, existential, but not problematic, possibility can be qualified as logical, physical, technical, psychological, economic, legal, etc. This is because it makes sense to say that, taking account only of the logical, the physical, the technical, the psychological, the economic or the legal aspect, it was possible for A to V; e.g. technically, but not legally, it is possible to print one's own bank notes. But one cannot say that, taking account of only of one of these aspects, it is possible that it is the case that A Ved, e.g. that A printed his own bank notes. The only aspect relevant to a possibility of something's being the case is the factual aspect, that is, whether there is any possibility that something is the case. Secondly, we can ask about existential possibility either 'How is it possible for A to V?' or 'Why do you think that it is possible for A to V?', but of problematic possibility only 'Why do you thing that it is possible that A V's?' and not 'How is it possible that A V's?', for the 'how' question asks for a method which enables A to V. Thirdly, the indistinctness of problematic, but not existential, possibility is sometimes characterized as 'slight', 'great', 'faint' or 'vague'. Fourthly, if it is possible that X Vs, then it follows that it is possible for X to V; but it can be possible for X to V without its being possible that X Vs. For instance, unless it is possible for me to read a numberplate at 30 yards, it is not possible that I did read one at that distance; but the fact that it is possible for me to do or have done it does not entail that it is possible that I do or did it. Having ascertained that it was undoubtedly possible for anyone in the office to have taken the money, the question arises about what actually happened. Is it possible that Bloggs took it? In other words, the possibility of an actuality implies, but is not implied by, the actuality of a possibility. Fifthly, existential possibility contrasts with necessity,[13] while problematic possibility contrasts with certainty and probability. 'It is possible that X Vs' is paralleled by 'It is probable that' and 'It is certain that' with the indicative, but there are no parallel forms 'probable to' and 'probable for', nor 'certain to' and 'certain for', to go with 'possible for X to V'. Nor is there any 'probable that' or 'certain that' with the subjunctive to correspond with 'possible that X should V', the variation of 'possible for X to V'. The legitimate form 'X is certain to V' is not parallel to 'It is possible for X to V'. It is equivalent to 'It is certain that X will V' and parallel to 'It is possible that X will V'. The fact that we can qualify the modal and ask '*How* is it possible to V?', but not qualify the modal and ask 'How is it necessary to V?' does not throw any doubt[14] on the connection between existential possibility and necessity. The reason for the difference is that to say that it is possible to V is to say that there is a way to V and we can, therefore, ask what this way is; whereas, to say that it is necessary to V is to say that there is no way not to V and we can, therefore, ask why there is no way. We can, of course, in both cases ask 'how' if we are only qualifying the main verb 'V 'in order to inquire about the manner of Ving.

A modal logic can be formulated for possibilty and certainty corresponding to that for possibility and necessity. Corresponding to the formula equating 'p is existentially possible' and 'It is not necessary that not–p'—for example, $Mp = NLNp$—would be a formula equating 'p is problematically possible' and 'It is not certain that not–p'—for example, $M_1p = NCNp$ (using 'M_1' for 'problematically possible' and 'C' for 'certain'). Conversely, the analogue of the modal formula '$Lp = NMNp$', which states that 'p is necessary' is equivalent to 'not–p is not possible', would be '$Cp = NM_1Np$', which states that 'p is certain' is equivalent to 'It is not possible that not–p'.

The logical dichotomy of existential and problematic possibility is revealed equally clearly in English by the other grammatical forms of 'possible', whether nominal, adverbial, or adjectival. Here again a clue to the difference is the presence of either a contrast with 'necessary' or a contrast with 'certain' ('probable').
(i) As regards the noun 'possibility', we can contrast the possibility *of* X (e.g. of knowledge, agreement, misunderstanding, fine weather, further play), *of* doing Y (e.g. going, staying, arriving) and of Z's happening (the house's falling, the weather's improving) either with the necessity or with the probability of any of these; though the possibility *that* X is Y contrasts only with the probability that X is Y and not with any necessity, as is sometimes made clear by the redundant form 'the possibility that X

may (might) be Y'. The possibility of so-and-so, like the possibility that so-and-so, can be slight, serious, good, faint, fair or definite. It can be considered or denied, met or avoided, discovered or overlooked, anticipated or ruled out.

(ii) The adverb 'possibly' seems to be almost invariably an indicator of problematic possibility. Here it is an idiomatic substitute for 'possible that' + the indicative, as in 'He possibly went by train' instead of 'It is possible that he went by train'. This is shown, first, by the fact that 'perhaps', 'probably' and 'certainly' are always possible grammatical alternatives, whereas 'necessarily' only rarely is in the affirmative form. In the negative form, the difference is clearer still, since 'possibly', like 'probably' and 'certainly', almost always precedes the negative, as it often does the affirmative, whereas 'necessarily' usually follows the negative. For example, contrast 'He possibly (probably, certainly) did not believe what you said' with 'He did not necessarily believe what you said'. The rarer form ' He necessarily did not believe what you said' would mean that his disbelief was necessary, as contrasted with the common form which means that his belief was not necessary. The exceptional double position of 'possibly' with 'could not' confirms all this. 'He possibly (probably, certainly) could not escape' means 'It is possible (probable, certain) that he could not escape', whereas 'He could not possibly escape' means 'It was not at all possible for him to escape'. 'Possibly' in the latter and in a few other forms, e.g. 'Could you possibly do it now?' is an intensifier.[15] There is no grammatical form 'He could not probably (certainly) escape' or 'Could you probably (certainly) do it now?'. Another indication that 'He possibly Ved' usually expresses problematic possibility is that it is quite appropriate to say 'He could have Ved, but he certainly did not', whereas it is absurd to say either 'He possibly Ved, though he certainly did not' or 'He may have Ved, though he certainly did not'.

A reason why 'possibly' does not usually occur with the meaning 'possible to' is that it would be redundant. Thus, 'It possibly follows' would only mean 'It follows, as is possible for it to do' on the analogy of the non-redundant 'It necessarily follows', which means 'It follows as is necessary for it to do'.

(iii) The adjective 'possible' can qualify, either existentially or problematically, only a limited set of nouns simpliciter;[16] namely those which indicate what something else either can or may—that is, what it is possible for it to or possible that it does—serve as, amount to or, in certain circumstances, be; or, perhaps, any noun when it has this function. Thus, we can say of something that it is *a* possible advantage, alternative, candidate, compromise, course, exception, exit, defect, hiding place, hypothesis, improvement, method, objection, opponent, position, result, sign, snag, source of misunderstanding, solution, precaution, reason, title, *the* possible result, author, cause, explanation or possible ruin, death, destruction. But not usually that something is *a* possible man, dog, tree, sea, sky, owl in a tree, flat tyre, etc., unless it is considered as something else that could or might also be this for a particular purpose, e.g. a tree for my garden or a sky for night flying.

Many of the nouns qualifiable by 'possible' are verbal nouns and, therefore, apply, if and only if some corresponding verb applies in the sense that advantages help, opponents oppose and signs signify. This grammatical feature is a consequence of the logical fact that Y is a possible X if and only if either it is possible for Y to or possible that Y does serve as or amount to an X (e.g. advantage, opponent, sign) and, therefore can or may do what an X does (e.g. help, oppose, signify). As signs signify and advantages help, so possible signs can or may signify and possible advantages can or may help. When the qualified X is not a verbal noun, e.g. train, day, school, target, coin, then something is a possible X only for something further. Thus, the coin is a possible coin for my collection, this is a possible school for my son, next Monday is a possible (the most likely) day for the invasion, two hundred cars a week is a possible (the firm's probable) target. Any verb in the offing here will indicate what is to be done to meet the further point, e.g. the coin can or may be added to my collection, the day can or may be that on which the invasion will take place.

Because, as we saw, 'Z is possibly an X' normally indicates problematic, not existential, possibility, 'Z is a possible X' implies 'Z is possibly an X' only where the former is problematic. Thus, if Z is a possible (probable) hiding place of the gang, then Z is possibly (probably) their hiding place; but if Z is a possible (necessary) hiding place for the gang, then it does not follow that

Z is possibly their hiding place. 'Z is possibly an X' does not, on the other hand, imply 'Z is a possible X' in either sense, since 'Z is possibly an X' is sensible whatever the X, whereas 'Z is a possible X' often, as we have seen, does not make sense. Of course it could not be that Z is possibly an X unless it were possible for Z to be an X; but to say that it is possible for Z to be an X, e.g. a hiding place (or an owl in a tree) does not mean that Z is a possible X. 'It is possible for Z to be an X' means it could be that Z is an X, whereas 'Z is a possible X' means 'Z could be an X', when X is something which various things other than Z could be.

The fact that 'Z is a possible X', 'Z is possibly an X' and 'There is a possibility of Z's being an X' can all be interpreted in terms of 'It is possible either that Z is an X or for Z to be an X' helps to make it clear that none of these implies that Z is an X or that there is any X.[17] Hence, a 'possible' X—like an 'alleged', 'intended' or 'appropriate' X and unlike a 'dangerous', 'brilliant' or 'multicoloured' X—is not a kind of X and, therefore, not a mysterious kind of X.[18] It is what could or might be an X even when it is not an X. Possibilities, possible occurrences and possible experiences are neither strange inhabitants of our familiar world nor familiar inhabitants of strange possible worlds. The present fashionable talk of what is possible as being what exists in possible worlds has all the vices of the once fashionable talk of what is only imaginary or apparent as being what exists only in our imagination or as an appearance.

An apparent exception[19] to the rule that 'Z is a possible X' does not imply 'Z is an X' underlines the rule. Some nouns have what could be called a misleading ambiguity, e.g. 'objection', 'explanation', 'solution', 'answer', 'course of action', 'choice', 'alternative'. On the one hand, X is often called an objection, explanation, solution, etc., if it is or could be put forward as one; on the other hand, it is sometimes only called such if it is proper, valid, legitimate, correct, etc. Similarly, Y is often called a choice, course of action or alternative, if it is or could be put forward as one; on the other hand, it is sometimes only called such if it is actually chosen, done or taken. It is in such cases that it looks plausible to hold that 'a possible X' implies 'an X'. A possible objection, explanation, answer, choice, course of action or alternative may be, on the one hand, a possible offered one or a possible correct or taken one. In neither case does 'a possible X' imply 'X', since a possible offered X need not be an actually offered X nor a possible correct or acted on X be an actual correct or acted on X. The belief that a possible explanation, solution, objection or choice, must be an actual explanation, solution, objection or choice depends on a shift between the elliptical and the non-elliptical uses of these words; for it means that a possible correct explanation, solution, etc., must be an offerable explanation, solution, etc. In short, there seems to be no genuine instance where 'It is a possible X' implies 'It is an X'.

A root cause of scepticism is the failure to see that although problematic possibility implies existential possibility, the converse does not hold. Sceptical arguments based on the premise that it is possible for something to be otherwise than it appears to be provide no reason for the conclusion that it is possible that something is otherwise than it appears to be, much less for the conclusion that something actually is other than it appears to be.

Attention was first drawn to this fallacy of scepticism by G. E. Moore[20] when, in a criticism of Russell, he correctly argued that because the conclusion (C1) 'It is possible for an experience of a certain sort not to have been preceded by a certain other experience' does follow from the fact that (P) 'An experience of the former sort has sometimes not been preceded by an experience of the latter sort', it does not follow that (C2) 'It is possible that this occurrence of an experience of the former sort was not preceded by an experience of the latter sort'. Furthermore, he correctly pointed out that to confuse C1 and C2 is like arguing that 'Because it is possible for a human being to be female, and I am a human being; therefore, it is possible that I am female'. Finally, he rightly suggested that Russell's fallacy rested, partly at least, on a confusion of different uses of 'possible'.

Moore seems to me, however, to have been wrong in two respects in his diagnosis of Russell's fallacy. First, he was wrong to say that the meaning of 'possible' in C1 is such that C1 means 'Some experiences of a certain sort are not preceded by certain other experiences'. In fact, though this can be evidence for C1, it is not equivalent to C1. Secondly, as well as saying that at the root of the fallacy there was a confusion of different uses of the word 'possible', Moore also said that there was a confusion about different uses of the word 'may'. I shall argue in a later discussion of 'may' that Moore was mistaken on this second point.

Other critics of scepticism, however, have confused the fallacy of arguing from existential to problematic possibility with the fallacy of arguing from existential logical possibility to existential empirical possibility. Thus, N. Malcolm,[21] in describing the general sceptical position and the particular Russellian argument as a move from (1) 'It is logically possible that p' to (2) 'There is some possibility that p' and then to (3) 'It is not certain that not-p', characterizes the move from (1) to (2) as a move from 'logical' to 'empirical' possibility. But in fact there are two steps in the move from (1) to (2), namely, a move from 'logically possible to be' to 'empirically possible to be' and a move from 'what it is possible to be' to 'what it is possible that it is'. Nor do sceptics commonly rest their arguments merely on what it is logically possible to be, but usually on what it is empirically possible to be, e.g. deceived by one's senses, from which they wrongly move to what it is possible that it is. The move from 'logical' to 'empirical' possibility is an illegitimate move from one sort of 'possible to be' to another sort of 'possible to be'; whereas the move to scepticism is an illegitimate move from either sort of 'possible to be' to 'possible that it is'. Conversely, determinism sometimes moves from the impossibility that it will be otherwise—that is, the certainty that it will not be otherwise—to the impossibility for it to be otherwise.

The difference between existential possibility and problematic possibility, that is, between its being possible for X to V and its being possible that X does V, is commonly, as the Oxford English Dictionary points out, expressed in English by the difference between 'can' ('could') and 'may' ('might'). Determinists often suggest that the fact that A Ved is incompatible with the supposition that he could have not Ved, while sceptics suggest that the fact that A might (may) have not Ved is incompatible with the supposition that he did V or, at least, with the supposition that he can be known to have Ved.

Let us, therefore, turn to an examination of the modal verbs 'can' ('could') and 'may' ('might') and to the relation of one to the other.

NOTES

1. Cp. Moore (1959), pp. 226–51; (1962), pp. 184–88; Hacking (1967), pp. 143–68; and Ayers (1968), pp. 33 ff.; Malcolm (1942), pp. 28–9, is confused between 'It is still possible that we shall go to California'—which means that nothing is certain yet—and 'It is still possible for us to go to California'—which means that nothing is necessary yet. Similarly Moore (1912), pp. 219–20, uses both 'it is possible for us to choose' and 'it is possible that we shall choose' for the use of 'possible' which is linked to uncertainty. Because of this he wrongly allows that it *was* possible that we would have chosen differently from what we did, whereas all that he should allow is that it *was* possible for us to have chosen differently. A similar confusion seems to underlie Ayers' (1968), p. 35, assimilation of 'It was possible that it would rain' and 'It could have rained' as well as his belief (p. 39) that 'It once was possible that X happened' makes sense and could be true even if it is impossible for X to happen.
2. *Pace* Ayers (1968), p. 39.
3. *Pace* R. Taylor (1965), pp. xv–xvi.
4. For this latter, cp. R. Taylor (1957), pp. 11, 20.
5. Cp. Toulmin (1958), pp. 54–7; contrast Kneale (1949), pp. 9–10.
6. Aristotle, *De Interpretatione*, chs. 12 and 13, very carefully pointed out the difference between τὸ δυνατὸν εἶναι, τὸ δυνατὸν μὴ εἶναι and τὸ μὴ δυνατὸν εἶναι; c.p. his similar remarks on ἐνδέχεται in *Prior Analytics* I.xiii.
7. (1959), pp. 240–1. Moore said 'if two people say it (sc. 'It is possible that p is true') at the same time about the same proposition, it is perfectly possible that what one asserts should be true, and what the other asserts false', because he thought that the expression was 'really an expression whose meaning is relative to the person who uses it'.
8. E.g. von Wright (1951b), pp. 8–9.
9. E.g. O'Connor (1960), p. 20 footnote; cp. von Wright (1951b), p. 11; Strang, pp. 462–5.
10. Strang, p. 465.
11. Loeb translation (1938) by H. P. Cooke. The Loeb often wrongly translates the contrast of τὸ δυνατὸν μὴ εἶναι and τὸ μὴ δυνατὸν εἶναι as 'It may not be' and 'It can not be'.
12. E.g. Anscombe (1956), p. 5.
13. Contrast Hacking (1967), p. 151, who contrasts it with 'impermissible'. But the substitution of 'permissible' for 'possible' often makes no sense, as in 'It is possible for X to have happened or for A to have undergone such-and-such'.
14. As Harré supposes.
15. Cp. Greenbaum, § 5.3.1.6.
16. Hacking's *grammatical* explanation (1967), pp. 154–8, of this limitation on the occurrence of 'possible', namely that 'something is a

possible N if and only if there is an active verb form V such that the thing is an N if and only if it V', seems to fit only verbal nouns.

17. This is why the epithets of actual X's are not generally applicable to possible X's; cp. Black (1960), pp. 117–26. Possible hiding places are not big or small, draughty or dry, well-concealed or easily found hiding places, though these can be characteristics of anything which is a possible hiding place.
18. *Pace* Goodman, pp. 37–62.
19. Hacking, *op cit.*, p. 154, is misled here by such philosophical solecisms as 'possible proposition'.
20. (1959), pp. 219–21.
21. (1942), pp. 27–9.

Chapter Two

Can

(A) CAN, COULD AND COULD HAVE

(i) First, a look at the grammar of this modal. 'Can' is a possible auxiliary to almost any verb except some other modals. It signifies that it is *possible for* somebody or something *to V*, whether the possibility referred to is that of a particular occasion ('It can be in the bottom of the drawer') or of, e.g., an ability ('He can read Greek', 'The car can do 100 m.p.h.'), an accomplishment ('He can get a First', 'He can catch that fish'), a perceptual power ('He can see it'), a characteristic ('He can be most unpleasant', 'It can be dangerous'), a regulation ('He can receive visitors three times a month').[1]*

'Could', which is the past form of 'can', signifies what grammarians sometimes call 'remoteness in time' and 'remoteness in reality' of a possibility. It signifies either that it *was* possible or that it *would be* possible for somebody or something *to V*. Its use to signify remoteness in reality is more common and covers every instance of possibility. In certain conditions A could read Greek, be very pleasant, visit the zoo, be in London, overhear what was being said. The use of 'could' to signify remoteness in time is more confined. It covers the possibility of abilities, characteristics and perceptual powers—when A was young he could read Greek, when he was an adolescent he could be pretty unpleasant and when he was at the barber's yesterday he could hear the conversation in the next room. The present possibility of some occurrence in the past, however, is expressed, not by 'could', but by 'can have',[2] which means that it *is* possible for it *to have Ved*. Thus, we say 'He can have been in London yesterday', 'He can have caught that fish', 'It can't have rained

* References to this chapter begin on p. 45.

yesterday' and not 'He could be in London yesterday', 'He could catch that fish' and 'It could not rain yesterday'. The 'can have' form seems to occur more commonly in the negative, e.g. 'She can't have been all that beautiful', 'He can't have expressed himself very well'. To say, on the other hand, 'It could be that he was in London, that it did not rain yesterday and that he caught that fish' is to signify remoteness in reality, not in time.

'Could have', which is the perfect form of 'can', also signifies remoteness both in time[3] and in reality. It signifies either that it *was* possible for somebody or something *to have Ved* or that it *would have been* possible for him or it *to V*. But just as 'could' is used to signify remoteness in time only for a possibility other than that of a particular occurrence, so 'could have', on the contrary, signifies such remoteness only for the latter. Thus 'He could have read it in the newspaper, been very unpleasant, overheard what was said' refers to an occasion, while 'He could read Greek before he was 8, be very unpleasant, overhear what was said' refers to an ability, characteristic or power.[4] In signifying remoteness in reality, on the other hand, 'could have' is applicable to any instance of possibility. Thus, in certain conditions, he could have spoken Greek before he was 8, been of a pleasant disposition, overheard a particular conversation and accomplished lots of things in life.

The difference between 'could have' and 'can have' is that the former means 'It *was* possible for A *to have Ved*' (or 'It *would have been* possible for A *to V*'), while the latter means 'It *is* possible for A *to have Ved*'. Admitting that I did not V, I can sensibly add, perhaps by way of explanation or excuse, that I couldn't have Ved; but it would be queer to add that I can't have Ved, which normally comes up only when it is still an open question whether I did or did not V. Past circumstances show that A couldn't have Ved, but present circumstances that he can't have Ved. Furthermore, 'couldn't have' can refer to remoteness in both time and reality, but 'can't have' only to remoteness in time. Hence, we can say 'He couldn't have done it, even if he had been richer', but not 'He can't have done it, even if he had been richer'.

(ii) Secondly, a look at the relations between the possible and the actual.

To say that A can V, could V or could have Ved does not imply either that A does or did V or that A does not or did not V. A car which can do 100 m.p.h., a man who could jump 5 feet when young and an eavesdropper who could have overheard his neighbour talking may either often or never have done so. Hence, two common and contrasting philosophical assumptions are mistaken: the one that 'can' and 'could' imply 'does' or 'did', the other that 'could have' implies 'did not'.

It is so obvious that lots of things which can be are not that the suggestion that 'can' implies or suggests 'is' only gains a hold in those cases where the normal or even the only evidence that something can be is that it is: for example, that a particular person can be very unpleasant or the weather very uncertain. A special and common case of this confusion of evidence and implication occurs with the perceptual powers. Mill held that the only evidence that something is visible, that is, can be seen, is that it is seen and contemporary philosophers[5] and linguists[6] sometimes say that if A *can* perceive, that is, can see, hear, etc., something, then it follows that A does perceive it. But this is a mistake. Though A does not at the moment see anything, it can properly, and often truly, be said of him that he can see, e.g., a number-plate at 25 yards, Big Ben from his study window, and what is behind him if he turns round. Furthermore, in the particular case, even though it were always true that if A can at a given moment see a book in front of him, then he does see a book, the latter is not entailed by, much less equivalent to, the former. The evidence for a statement should not[7] be confused with the truth of the statement. The evidence that I can at this very moment reach the book in front of me may or may not consist in my actually reaching it, whereas the normal, perhaps the only, evidence that I can at this very moment see the book does consist in my actually looking and seeing. Similarly, the usual proof that I can remember, imagine or think of X, is my doing so. But it is because what A can do is not limited to any occasion on which he does it, that, although one says both 'He sees (saw) a series of flashes' and 'He can (could) see a series of flashes', one says 'He sees (saw) a sudden flash' but not normally 'He can (could) see a sudden flash'.

Similarly, 'He could jump a 5 foot fence when he was young'

would normally be asserted on the basis of the knowledge that he had done so; but there is nothing self-contradictory in my claim that I could do this even though I had never tried to do so. There are lots of things a good athlete could now do though he has not done them and lots of things I could at the moment be doing though I am not.

Hence, it is a mistake to suggest that where 'could'—or 'can' —is used of a particular action or accomplishment, then 'A could V'—or even 'A was able to V'—is 'properly said only when A has succeeded in Ving'[8] or that there is a sense of 'being able' which 'doubles for "managing" and "bringing off" '[9] or that the assertion that A could not V or that A was not able to V 'can only be correctly made if the agent has been unsuccessful despite the fact that he tried'. For, first, if 'A could V' implied 'A did V', then 'A did not V' would imply 'A could not V'. But, for example, 'He did not get the tickets' does not imply 'He could not (was not able to) get the tickets'; for there may be other reasons for his not getting them. Secondly, 'He could not go to the meeting', which certainly implies that he did not go, does not imply that he tried, unsuccessfully, to go. Thirdly, if 'was able' implied 'did', one would expect that 'is able' implied 'does'. But clearly it can be true that A is able to do something which he does not do or try to do. Fourthly, we can sensibly ask whether A could (or was able to) manage to do or bring off something, but it makes no sense to ask whether he could (or was able to) be able to do it.

The mistake in these suggestions, as in the others, lies in confining oneself to examples of accomplishments, e.g. 'see', 'overtake', 'sink', 'succeed', where the evidence for the applicability of 'could' (or 'was able') and 'could not' (or 'was not able') consists in successful or unsuccessful attempts. Statements about what someone was at a particular time able or unable to accomplish are often based on, but do not imply, what at the time he did or did not accomplish.

Although it does not follow that what can or could be is or was, it does follow that what is or was can or could be; a truth enshrined in medieval logic as 'Ab esse ad posse'. For, if the latter did not follow, the fact that something cannot or could not be so would not preclude that it is or was so. And that is absurd.

But to insist that *is* and *was* imply *can* and *could* is not to suggest that 'it follows merely from the premise that he does it, that he has the ability to do it, according to ordinary English';[10] for, as we have seen, 'can' does not mean 'has the ability'. What is done, much less what happens, need not be the exercise of an ability. Hence, conversely, to hold that 'He has the ability to do it' does not follow from 'He did it' is not to impugn the implication of 'can' from 'do'. [11]

When what is asserted is that A *could have* Ved, the line commonly taken by philosophers[12] is that this suggests, even if it does not imply, either that A did *not* V or that the speaker supposes that A did *not* V. This line is linked to the false belief that 'could have' is used only in the apodosis of an unfulfilled conditional, as in 'If it had not snowed, they could have got through the pass'. But this view is doubly mistaken. For, first, 'could have' may signify an actual past since it covers remoteness in time as well as in reality and, secondly, it can be used even when the speaker suspects that A did V. Thus, 'He could have got in by the back door' need not either be hypothetical or suggest that he did not get in that way or that the speaker thinks he did not. Underlying these false theses, is, once again, the plausible assumption that one would not normally say that A could have Ved if one knew that A did V. Hence 'A could have Ved' is usually said either in ignorance of the truth of 'A did V' or, what is different, in the knowledge of the truth of 'A did not V'.

When 'can' is used in the asking for or the giving of permission, e.g. 'Can I go now?', 'You can', it will have certain grammatical and logical peculiarities. For instance, there is no permissive use of 'can have' or 'could have' because permission contains no reference to the past. 'Could' is used in tentative requests for and reports of, but not in the granting of, permission. 'You can do it' can be qualified as a statement in ways in which it cannot be qualified as permission, e.g. 'You can easily, probably, do it'. But 'can' still retains in permission the same sense as it does elsewhere, namely that it is possible to V. The difference between the possibility of permission and that of ability, opportunity, accomplishment, circumstances, etc., is solely in the different kinds of grounds for the possibility. In asking for or granting permission what is at issue is whether someone's wishes make it impossible for another to V.

(B) COULD AND WOULD

Philosophers have recently devoted much attention to the relations between *could* and *would* because it has seemed plausible to many that 'I could have done otherwise', which is a normal expression of our belief in free will, is elliptical for 'I could have, if I had chosen (wanted)' and that either the original 'I could have' or this 'I could have, if' is to be analysed as 'I *would* have, if I had chosen (wanted)'. Let us, therefore, look at the actual behaviour of *could* and *would*.

We saw that 'could', which is the past form of 'can', and its perfect 'could have' signify remoteness either in time or in reality. To say of someone or something that it could V (or could have Ved) is to say either that it was possible or that it would be possible for it to V (or to have Ved). We saw also that the use of 'could' and of 'could have' to signify remoteness in reality is more common than its use to signify remoteness in past time, where 'could' is confined to the possibility of abilities, characteristics, perceptual powers, etc., and 'could have' to particular happenings, actions or situations.

'Would', which is a past form of 'will', and its perfect 'would have' also signify either remoteness in time, e.g. 'He would not accede to their demands', 'He would not have gone via Rome', or in reality, 'If he went by air, he would be there now', 'If he went by boat, he would not have reached New York yet'.

It is clear that in general 'would' (or 'would have') and 'could' (or 'could have') are quite different.[13] The answers to the question 'What *could* he have done with the money?' are more extensive than, and indeed include, the answer to the question 'What *would* he have done with the money?', whether or not we add the supposition 'if he had only a week to spend it in'. 'Could have' indicates the possibilities which were or would have been open to him, while 'would have' asserts which one of them he would have taken.

The if-clause in 'could have if' indicates what would have

made something possible, while the if-clause in 'would have if' indicates what would have made it actual. 'Would have' implies, but is not implied by, 'could have'; and 'I would if I could' is certainly not a tautology, equivalent either to 'I would if I would' or to 'I could if I could'.

The difference between 'could have if' and 'would have if' still persists when the if-clause includes a volitional verb, such as 'choose', 'want', 'will'. 'Could I have gone by air if I had wanted to?' is a question which I could properly ask of the travel agency; but 'Would I have gone by air, if I had wanted to?' is a question about my own psychology. The latter question, but not the former, is an inquiry about a causal connection. 'He could have killed you, if he had wanted to' indicates his powers and opportunities; 'He would have killed you, if he had wanted to' indicates his character. The assurance given in 'If he had wanted to come, he would have' is quite different from that given in 'If he had wanted to come, he could have'.

The difference between 'could have if' and 'would have if', when the if-clause contains a volitional verb, comes out in the contrast between the force of 'if' in the two cases. The 'if' of both 'could have if' and 'would have if' is implicative, that is, allows for contraposition, when the if-clause contains a non-volitional verb. Both in 'If he had been there, I *could* have asked him' and in 'If he had been there, I *would* have asked him' the falsity of the apodosis entails the falsity of the protasis. In both also the if-clause introduces a condition. When, however, the if-clause contains a volitional verb, the 'if' of 'would have if' remains implicative, but the 'if' of 'could have if' often does not. Thus, granted the truth of 'If I had wanted your help, I would have asked for it', the fact that I did not ask for it shows that I did not want it; it's because I did not want it that I did not ask for it. The original statement, indeed, asserts that my asking for your help was conditional on my wanting it. I could in this way be contrasted with a proud man of whom it might be said 'He would not ask for your help, even if he wanted it'. On the other hand, we cannot argue that, granted the truth of 'If I had wanted your help, I could have asked for it', the falsity of 'I could have asked for it' would show that I did not want it. For, the original statement does not assert that the possibility of asking for your help was conditional upon wanting it; it asserts that the possibility was there, to avail myself of if I wanted it. Therefore, the supposition of the falsity of 'I could have asked for it' is inconsistent with what was originally granted. My position here could be contrasted with that of a mute man of whom it might be said 'He could not ask for your help, even if he wanted it'. 'He could have done anything he wanted (chose, liked)', which can be regarded as a generalization of 'He could have done X, if he had wanted (chosen, liked) to do X', defines the range of what he could have done rather than the conditions which made it possible for him to do it. In other words, the logical parsing is 'He could have done (anything he wanted)' and 'He could have done (X if he had wanted)' rather than 'He could have done anything (if he had wanted)' and 'He could have done X (if he had wanted)'.[14]

The behaviour of the other modals in relation to 'if' clauses containing a volitional verb is rather different from that of 'can'. First, with 'can' the main verb of the protasis and the apodosis is usually the same, i.e. 'I can V, if I want (choose) to V', for the question is whether I can do what I want (choose) to. Thus, we say 'If I want (choose) to go to the University, I can 'go', but not 'If I want (choose) to go to the University, I can have two A-levels'. With the other modals the two main verbs are different, i.e. 'I ought to (need to, must) V, if I want (choose) to F', for there is no plausibility in the thesis that I ought to (need to, must) do what I want (choose) to. Thus, we say 'If I want (choose) to go to the University, I ought to (need to, must) have two A-levels', but not 'If I want (choose) to go to the University, I ought to (need to, must) go'. An example like 'If you want (choose) to make a noise, you ought to (need to, must) make it elsewhere' is only an apparent exception, for making a noise elsewhere is contrasted with making it here.

Secondly, what I ought to (need to, must) do is frequently a means to, or a method of, doing what I want (choose)—though not, of course, of wanting (choosing) it—but what I can do is not. It is because I want (choose) to go to the University that I ought to (need to, must) have two A-levels, but it cannot be because I want (choose) to go that I can have them. Wanting (choosing) to go to the University makes it appropriate (or necessary), but it does not make it possible, to have two A-levels.

Thirdly, and consequently, the 'if' of 'ought to (need to, must), if', unlike that of 'can, if' is implicative even when the if-clause contains a volitional verb. Hence, 'If I want (choose) to go to the University, I ought to (need to, must) have two A-levels' has a valid contrapositive form. If a student tells me that he does not need two A-levels for what he wants (chooses), then I can conclude that what he wants (chooses) cannot be to go to the University.

Although the 'if' of 'I could have, if I had wanted (or chosen)' is usually non-implicative, it is not, I think, always so. Consider this example. Just as some men are so constituted that in some circumstances it can truthfully be said of them that they *do* what and only what they want, so they could be so constituted that in some circumstances it could be truthfully be said of them that they *can do* what and only what they want (or choose). Of the former sort of man it would be true to say both 'If he wanted (chose) to do X, he would do it' and 'If he did not want (choose) to do X, he would not do it'; while of the latter it would be true to say both that 'If he wanted (chose) to do X, he could do it' and 'If he did not want (choose) to do X, he could not do it'. My claim that I couldn't have made Celia happy can sometimes properly be met with your reproach that I could have done it if I had really wanted to or wanted to enough. And, conversely, your reproach that I could have married Anna can sometimes properly be met with my claim that I could not marry someone whom I had not chosen and did not want. In other words, there do seem to be cases when the 'if' of 'A could have if he had wanted (chosen)' and of 'A could not, if he had not wanted (chosen)' are implicative and where wanting or choosing to do so-and-so can be either a sufficient or a necessary condition for doing something. These are cases where wanting and choosing come very close to trying and making an effort of will.[15] Here we appreciate the force of the sayings 'You can do anything, if you really want to (want it enough)' and 'If my heart's not in it, I can't do it'. Other cases might be found amongst addictions.

My conclusion, therefore, is that the 'if' of 'would have if' is always implicative whatever the verb of the if-clause; whereas the 'if' of 'could have if' is often, but not always, non-implicative when the if-clause contains a volitional verb.

Now both these points have been disputed. Austin,[16] on the one hand, seems to deny not only that the 'if' of 'could have, if he had chosen' is ever implicative, but also that the 'if' of 'would have, if he had chosen' is any more implicative than the 'if' of 'could have, if he had chosen'. Moore and other philosophers, on the other hand, have asserted not only that the 'if' of 'would have, if he had chosen', but also the 'if' of 'could have, if he had chosen', is always implicative.

Consider first Austin's doubts[17] about the implicative nature of the 'if' of 'would have, if he had wanted (chosen)'. He admits (p. 159) that it is unlike the 'if' of '*could* have, if he had wanted (chosen)', since it is never permissible to argue from 'I *would* have, if I had wanted (chosen)' either to 'I would have, whether I wanted (chose) to or not' or simply to 'I would have' in the way that it is, at least sometimes, permissible to argue from 'I *could* have, if I had wanted (chosen)' to 'I could have, whether I had wanted (chosen) to or not' and to 'I could have'. On the other hand, to the query whether we can argue by contraposition from 'I would have, if I had wanted (chosen)' to 'if I did not, I did not want (choose) to'—which would prove the 'if' to be implicative—he answers rather hesitantly (p. 159; less hesitantly on pp. 157 and 165) 'I think not', while insisting that 'it would still be patently wrong to conclude that the meaning of "I would if I had wanted (chosen)" is that my choosing to do the thing was sufficient to cause me inevitably to do it or had as a consequence that I would do it'.[18] Now both the hesitant and the insistent answers must be mistaken in the light of my earlier example—namely, If I had wanted your advice, I would have asked for it, therefore the fact that I did not ask for it is due to my not wanting it—which showed the 'if' of 'would if I had wanted (chosen)' to be both implicative and conditional. The main source, I think, of Austin's hesitancy about the implicative nature of the 'if' here is his reliance on an example using the first person present tense of the verb, that is 'I shall if I choose'. Because he relies on this example he can contrast the 'shall' of 'I shall if I choose' with the 'shall' of 'I shall ruin him if I am extravagant'; first, as 'that mysterious old verb shall' in contrast to the ordinary auxiliary 'shall' and, second, as an 'expression of intention' in contrast to an 'assertion of fact', and conclude that the 'if' of 'I shall, if I choose' is the 'if of stipulation'. Unfortunately, however, an example using the first person past tense, that is 'I would have, if I had wanted (chosen)', employs neither a 'mysterious old verb shall' nor plays the role of 'an expression of intention'.[19]

I conclude, therefore, that the 'if' of 'I would have, if I had wanted (chosen)' is implicative and even conditional; that, e.g., if it is true that I would have married her, if I had wanted (chosen) to, then my failure to marry her is due to my not wanting (choosing) to.

Consider, secondly, the contradictory opinions about the 'if' of '*could* have, if he had wanted (chosen)'; namely, that it is never implicative (e.g. Austin) and that it is always implicative (e.g. Moore). If the examples I have given are correct, then both these extreme theses are mistaken, for the 'if' in this case is usually, but not always, non-implicative.

The thesis that the 'if' of 'I could have, if I had wanted (chosen)' is always implicative has usually been simply taken for granted[20] mainly because philosophers have quickly turned their attention to the alleged analysans of 'could have', namely 'I would have, if I had wanted to', whose 'if' can more plausibly be considered implicative and even conditional.[21] Such a diversion of attention lies also at the basis of the suggestion[22] that the 'if' gives a sufficient condition, not for the existence of a possibility, but for its realization.

The thesis that the 'if' of 'could have, if he had wanted' is always non-implicative and that the if-clause never gives a condition for the truth of the 'could' clause is held by nearly every philosopher who has recently examined the problem.[23] I have argued earlier that, with the possible exception of what can be called the 'will-power cases', this is normally so.

We are now in a position to consider several theses which have been advanced in recent years about the relation of 'could' and 'would'. It has been suggested (a) that 'could have' is elliptical for 'could have if';[24] (b1) that 'could have if' is to be analysed as 'would have if';[25] and (b2) that 'could have' itself is to be analysed as 'would have if'.[26] In all the cases different views have been expressed about what follows the 'if'. Sometimes (e.g. Nowell-Smith, Thalberg, Ayers) it is claimed that there is an indefinite set of conditions which vary with what verb accompanies 'could have'; but usually the claim has been either that what follows 'if' is a volitional verb, such as 'chooses', 'wants', 'likes', 'wills' (e.g. Moore, Baier, Ewing), or that what follows 'if' is the verb 'try' (e.g. Chisholm, Honoré, Dore). And, as we have seen, different views have been expressed about the nature of the clause introduced by 'if'. Is it different for 'could if' and for 'would if'? Does it give in either case a sufficient condition, a necessary condition, or not a condition at all?

(a) Consider first the suggestion that 'could have' is always elliptical for 'could have if'. Although Moore said (1912, p. 211) that it was a 'very obvious suggestion' that what is meant by 'I could' is 'I could if I had chosen' and that he himself had used the former as an abbreviation of the latter in the early part of his book, he did not give any proof that this abbreviation occurs in ordinary language; nor did he think it important to distinguish it in detail from the quite different suggestion that 'I could' is a 'short way of saying' 'I should, if I had chosen'.

There is in fact no warrant in grammar[27] or ordinary use[28] for not distinguishing the unconditional use of 'could have' and the conditional use of 'could have if', either for 'if I had chosen' or for any other if-clause. 'Could have' is equivalent to 'it was possible to' while 'could have if' is equivalent to 'it would have been possible to . . ., if . . .'. As regards non-volitional if-clauses, 'He could have done it, if he had had more time', for example, clearly differs in sense from 'He could have done it in the time available'. The drawback of a volitional if-clause, on the other hand, is that 'if he had chosen' is not a meaningful addition to 'could have' statements about inanimate objects, e.g. 'The oil could have drained out of either plug'.[29] Furthermore, the suggestion that 'could have' is short for 'could have if' would both lead to an infinite regress and also make nonsense of *modus ponens.* It would lead to an infinite regress,[30] since the 'could have' part of 'could have if' would itself have to be expanded to 'could have if' and so *ad infinitum.* It would make nonsense of *modus ponens,*[31] for we could not argue 'I could have, if X; But X; Therefore, I could have', since, according to this view, 'could have' would not have an unconditional use. Further, the fact that I did X is sufficient to establish that I could have done

X; no proof that I wanted, chose or tried, to do X is necessary.[32] Further, the truth of 'He could have done it, if he had wanted (chosen)' and the truth of 'He could have done it, if circumstances had been different' are compatible with the falsity of 'He could have done it',[33] particularly in the case where he did not or could not want (or choose) to do it[34] or where the circumstances were not different. Finally, 'could have' is not elliptical for 'could have, if I had wanted'; the latter is commonly a logically superfluous version of the former aimed at emphasizing that in the circumstances the 'could have' is quite unconditional, e.g. 'I could have had you killed, dismissed, transferred . . . if I had wanted', 'I could have gone by air, by road, or a later train . . . if I had wanted'.

(b1) Consider next the suggestion that '*could* have if' is to be analysed as '*would* have if'. In putting forward his suggestion that 'could have' is short for 'could have, if I had chosen', Moore added (1912, p. 211) 'or (to avoid a possible complication) perhaps we had better say "that I *should if* I had chosen" '. Critics have differed whether Moore, therefore, intended to assert that 'could have if' and 'would have if'—at least where the 'if-clause' is 'if I had chosen—are equivalent (e.g. Austin) or not (e.g. Ewing). The exegetical point, however, is unimportant, since some philosophers[35] have held not merely that Moore did intend to assert this, but also that he was, or would have been, correct to do so.

Since I have argued that 'could have if' cannot be equivalent in meaning to 'would have if', whatever the if-clause, it is only necessary to examine any explicit argument which philosophers have given for its alleged equivalence. So far as I know, the equivalence has been alleged only where the if-clause is volitional.[36]

Since 'I would have done it, if I had wanted (chosen)' certainly entails 'I could have done it, if I had wanted (chosen)'[37]—because 'I would have' entails 'I could have'—the questions are (i) whether 'I could have done it, if I had wanted (chosen)' entails 'I would have done it if I had wanted (chosen)'; (ii) whether a mutual entailment of 'I could have done it, if I had wanted (chosen)' and 'I would have done it, if I had wanted (chosen)' implies equivalence in meaning; (iii) whether, if it does not, there are any other reasons for supposing equivalence in

meaning. As regards (iii), I know of no such reason. As regards (ii), it is generally accepted that logical equivalence is a necessary but not a sufficient condition for equivalence in meaning. Some supporters[38] of the logical equivalence of these two propositions do, indeed, admit a difference in meaning, though others[39] assume the logical equivalence is equivalence in meaning. As regards (i), the position is more difficult. It has been stated[40] that ' "I could have done it if I had chosen" would ordinarily be understood as entailing that "I should have done it if I had chosen", since the phrase implies the absence of fatal external obstacles'. It has also been claimed[41] that it would be self-contradictory to affirm the former and deny the latter. But such an unargued assertion seems no more plausible[42] than the corresponding assertion, to be discussed below, that 'I could have done it' itself entails 'I would have done it, if I had wanted'. Indeed, it is less plausible since 'could' itself certainly does not mean 'would' and, therefore, it is unlikely that 'could if' means or entails 'would if', though possible that 'could' itself means or entails 'would if'. Furthermore, if we stick to 'want' as the volitional verb in the if-clause, it is easy to find occasions on which we certainly do not use 'could if' to imply 'would if'. For instance, it may well be true that I could have accepted that invitation last week if I had wanted to and I wanted very much to accept, but I did not accept because I thought, for various reasons, that I ought not to; but it could not be true that I would have accepted it if I had wanted to and did want to, but did not accept. Admittedly, such an example is more doubtful for 'could if I chose' since it is possible to disallow that I chose to do something but did not do it. Since, however, there seems to be something equally queer about 'I chose to do it but I could not do it', this suggests that the idea of 'I can if I *choose*', unlike 'I can if I *want*', needs more cautious examination that it has been given or can be given here. Finally, it is clear that the permissive 'You can, if you choose (want)' does not imply 'You will, if you choose (want)'.
(b2) Let us, therefore, consider finally the suggestion that 'could have' itself has to be analysed as 'would have if'. The suggestion has been adopted by many philosophers,[43] though they have differed on the content of the if-clause—whether it should contain a volitional verb, such as 'want' or 'choose' (e.g. Moore,

Ewing), the verb 'try' (e.g. Honoré, Chisholm) or an unspecified and various set of conditions (e.g. Nowell-Smith, Ayers). Although the supporters of these analyses agree that the if-clause is implicative, some hold it is causal (e.g. Moore, Ewing) and others not (e.g. Nowell-Smith, Taylor).

There are, however, various general objections to any analysis of 'could have' in terms of 'would have if' and particular objections to particular forms of it.

First, all three variations of this analysis are based on the supposition that 'could have' signifies a power or disposition in an agent or object to accomplish something, e.g. an ability to read or the capability of burning or heating (e.g. Nowell-Smith, Ewing, Taylor, Thalberg, Ayers, Pears). In other words, that 'could have' is akin to 'was able to'. Hence, the plausibility of analysing a power or disposition in conditional terms is transferred to the analysis of *could have*. Such a narrowing of the problem is natural enough for philosophers primarily interested in free will and human behaviour. But 'could have' often operates where abilities and capabilities are absent. There are many sorts of reasons why it is possible for X or P to V which are unconnected with any ability or capability in X or P. This is clear from the fact that 'could have' is often unrelated to 'was able'. The possibility may be due to a liability in X or P, e.g. 'The tree could have fallen either way', 'The inspector could have been mistaken'; or to a rule, e.g. 'The letter could have been sent either by registered post or recorded delivery', 'The candidate could have sat a different paper'; or to circumstances, e.g. 'He could not have been present, because he was not born then'. In none of these cases would it make sense to analyse 'could have' as 'would have if', whatever the content of the if-clause. What are the terms of the 'would have if' translation for 'could have been mistaken' or 'could have fallen'?

A common philosophical move[44] to avoid this objection is to argue that 'could have' has different senses when used of abilities, capabilities, liabilities, circumstances, rules, etc. But this is a confusion of meaning and grounds, which becomes clearer when 'could have' is translated into 'possible to'. The move[45] now takes the form that 'psychologically possible', 'physically or causally possible', 'logically possible', 'legally possible', etc., are all different senses of 'possible'. But they are not. The adverbial qualification does not introduce different senses of 'possible'; it introduces different grounds for the possibility or different aspects under which it is considered. This mistake is exactly the same as that often made by philosophers who call, e.g., 'legally obliged', 'morally obliged', 'logically obliged', and 'physically obliged', different senses of 'obliged'; or 'legally responsible', 'morally responsible' and 'causally responsible' different senses of 'responsible'. But 'legally responsible' and 'morally responsible' are no more different senses of 'responsible' than 'officially in charge', 'nominally in charge' and 'temporarily in charge' are different senses of 'in charge'. It is a mistake to assume that because 'Mly Q' and 'Nly Q' have a different meaning, therefore, 'Q' has a different meaning when qualified by 'Mly' from what it has when qualified by 'Nly'. On the contrary, the Q which 'Mly' qualifies is the same Q which 'Nly' qualifies.

I conclude, therefore, that the plausibility of the 'would have if' analysis of 'could have' rests largely on a confusion with a 'would have if' analysis of powers and dispositions.[46]

Secondly, the particular 'would have if' analyses which introduce either a volitional verb, e.g., 'want' or 'choose' or the verb 'try' into the if-clause have the fourfold defect that, first, they are only plausible when 'could have' is restricted to animate subjects; second, that, even there, they give for some verbs a curious tautology; third, that in some substitutions they are mistaken; and, fourth, that they are often senseless.

As regards the first defect: how can 'The statement could have been false' or 'The bottle could have been broken' be analysed in terms of hypothetical volitions or attempts? Nor will it do to suggest, as does Austin, that 'could have' with an inanimate subject is a vulgarism for "might" have'. 'Might have', as we shall see, is logically quite different from 'could have'. And even if we did substitute 'might have' for 'could have' the objection would still stand, since 'might have' implies 'could have' and would, therefore, on this thesis imply the absurd 'would have if'.

The second defect arises when what is taken as an example of 'could have Ved' is 'could have *wanted*, *chosen* or *tried*'. For the analysis of this has to be of the form 'would have wanted (chosen or tried) if he had wanted to want (chosen to choose, or tried to try)'; none of which seems intelligible.[47] The third defect is that some substitutions of the suggested analysans for the analysandum lead to error. Thus, to replace 'He could have Ved' by 'He could have Ved, if he had wanted to V' in 'He could have Ved, if he had wanted to V' gives 'He would have Ved, if he had wanted to V, if he had wanted to V'. But this is either meaningless[48] or is equivalent[49] to the simpler 'He would have Ved, if he had wanted to V'. It follows, however, from the latter supposition that 'He could have, if he had wanted' is equivalent to 'He would have, if he had wanted'. But we have already seen that this is mistaken.

As regards the fourth defect, volitional and attempt verbs often make no sense even where the subject of the 'could have' clause is animate or personal.[50] This is obvious with verbs in the passive voice, e.g. 'He could easily have been drowned' or 'He could not have been dead more than five minutes'. But it is also true of many verbs, e.g. 'He could have made a mistake about X, known Y, been expecting or hoping for Z'. Indeed, sometimes where a volitional verb makes sense, 'try' does not; e.g. contrast 'He would have refused, if he had wanted' with 'He would have refused, if he had tried'. Nor can this sort of objection be met[51] by changing 'He would have done X, if he had tried to do X' to 'He would have done X, if he had tried to do Y' since it is difficult in the way just mentioned to see exactly what value could be substituted for 'Y' to give the same sense as the original or, indeed, any sense.

The type of 'would have if' analysis of 'could have' which leaves unspecified the content of the if-clause[52] might appear to escape these objections. It is difficult, however, to see what content it could give to the if-clause of 'would have if' other than a volitional or attempt verb which would not turn the unconditional 'could have' into a conditional 'could have'.

The third feature to emphasize about all 'would have if' analyses of 'could have' is their reliance on the alleged logical equivalence between 'could have' and 'would have if'. Three questions, therefore, arise: (1) Does 'could have' imply 'would have if' for some 'if'; (2) Does 'would have if' imply 'could have' for some 'if'; and (3) if 'could have' and 'would have if' are logically equivalent for some 'if', is this due to an equivalence in meaning?

Most philosophers who have held that 'could have' can be analysed as, or is equivalent in meaning to, 'would have if' have not bothered to distinguish between the thesis that these are logically equivalent, that is, are mutually implicative, and the thesis that they are equivalent in meaning.[53] Furthermore, they have given the same reasons for holding that 'could have' implies 'would have if' as for holding the converse.

By far the most common reason given is that 'could have' refers to an ability or capacity or even to a particular opportunity for their exercise, and is, therefore, equivalent to some form of 'would have if' (e.g. Nowell-Smith, Honoré). For this reason, even those who hold that 'could have' is not equivalent to 'would have if' nevertheless argue that the former implies the latter.[54] The objection to this is twofold. First, it may be disputed[55] whether abilities and capacities, or even dispositions, are correctly analysed in hypothetical terms, though it is difficult to prove that no 'would have if' is implied by a dispositional 'could have', especially when the 'would have if' is advanced in some weak form, such as 'would usually have, if[56] or where 'He could have done X' is analysed as 'There is some Y, such that if he had tried to do Y, he would have done X'.[57] The second, and more important, objection is that, as we have already seen, 'could have' does not always indicate an ability or capacity.

The only other reasons in support of a 'would have if' analysis of 'could have' with which I am familiar are the rather weak ones given by Moore (1912, pp. 212–17), namely, (1) that he 'cannot find any conclusive argument to the contrary'; and (2) that those who deny or produce evidence against the belief that men could have acted differently from what they did commonly do so by denying or producing evidence against the view that they would have acted differently if they had chosen.

Whether or not 'could have' implies 'would have, if', the latter does not imply, and therefore cannot be equivalent to, the former. For we cannot *deduce* that 'X could have happened' from the fact that 'X would have happened if Y had happened' where it is impossible for Y to have happened, e.g. that the metal could have melted does not follow from the fact that it would have melted if it had been heated to an impossibly high temperature. This is not, however, to say that it is *false* that 'X could have happened'.[58] This would be false only if the proposed analysandum had been 'X would have happened only if Y had happened'.

Another possible objection to the thesis that 'would have if' implies 'could have' is that 'would have' alone implies, though it is not implied by, 'could have'. 'Would have if' gives a hostage to fortune not given by 'would have' alone and, therefore, allows by virtue of an unfulfillable antecedent the just mentioned possibility not allowed by 'would have' alone, that the antecedent is true and the consequent false.

A recent objection to the analysis of 'could have' in terms of 'would have if' is this.[59] 'X *would* have happened (if and) only if Y had happened', but 'Y *could* not have happened'—or 'X *could* have happened (if and) only if Y had happened' but 'Y *did* not happen'—is inconsistent with the analysandum 'X could have happened' but not with the alleged analysans 'X would have happened, if Y had happened'. For example, George Washington was reputedly a pathologically honest man. Suppose, for the sake of argument, that he was also pathologically stubborn; that is, that he did what and only what he wanted (chose) to do. Then it could be true both that George Washington would tell a lie if and only if he wanted (chose) to, and that he never could want (choose) to tell a lie; thus, he never could tell a lie. Hence, to say that someone would have done X if he had wanted (chosen) to do it cannot be equivalent to, much less synonymous with, the statement that he could have done it.

Finally, even if 'could have' and 'would have if' were logically equivalent, this would be no proof that they were equivalent in meaning. Nor, indeed, are they. To say what could have happened or what someone could have done, is to mention the possibility open in the circumstances; to say what would have happened, or would have been done, if . . ., is to make a conditional prediction. The truth of such a conditional prediction may be evidence for, and even proof of, the truth of the possibility, but it is not the same as it.

(C) 'COULD HAVE DONE OTHERWISE'

An investigation of the modal concept expressed by 'could have' throws some light on the dispute between those who believe in Free Will and those who believe in Determinism as to whether anything other than what was done could have been done. We saw in the first section that 'A could have Ved' does not imply, or even suggest, either that A did V or that A did not V. Therefore, by contraposition, neither 'A did not V' nor, of course, 'A did V' implies that A could not have Ved. Hence, there can be no logical incompatibility between 'A did not V' and 'A could have Ved'. The assumption made by all of us in our non-philosophical moments, that of the things that did not happen some could have happened while others could not have happened, makes perfectly good sense. Furthermore, this makes good sense without supposing, what we saw in section (B) would be mistaken, that 'could have' is either elliptical for 'could have, if' or equivalent to 'would have, if'. To say that a car, which did not do 100 m.p.h., could have done 100 m.p.h., but not 200 m.p.h., is not to say that it *could* have done so, *if*, e.g., the carburettor had not been blocked. It is to state categorically that in the actual circumstances it could have done 100 m.p.h. Of course, if it *had* done 100 m.p.h., the circumstances would have been different; but the claim is only that in the given circumstances it *could* have done, not that it *did*, 100 m.p.h.

Nor is to say that the car *could* have done 100 m.p.h. to say that it *would* have done 100 m.p.h., *if*, e.g., the throttle had been fully open. The throttle's being fully open—which, of course, it was not—is a condition which, if it had been present, *would have made* the car *actually* do 100 m.p.h. It is neither a condition which *made it possible* for the car to do 100 m.p.h. nor a condition which—like unblocking the carburettor—*would have made it possible* for the car to do 100 m.p.h. An unconditional possibility ('could') must not be confused either with a conditional possibility ('could, if') or with a conditionally predicted actuality ('would, if'). Nor must the conditions necessary or sufficient for the actualization of a certain possibility ('would, if p') be confused with the conditions necessary or sufficient for the existence of that possibility ('could, if p'). Conditions which may have been necessary in order that what could have been so were so need not have been necessary in order that that could have been so. The actualization of a possibility normally requires more than the existence of that possibility just as the exercise of an ability may need a stimulus which the possession of that ability does not. Contrariwise, to prevent something from happening is not necessarily to prevent the possibility of its happening.

Part of the feeling that what was not could not have been is due to this mistaken supposition that a condition which is necessary for something's being so—e.g. the throttle's being fully open for the car to do 100 m.p.h.—is a condition which is necessary for its being possible for something to be so—e.g. the carburettor's being unblocked for its being possible for the car to do 100 m.p.h. But just as the difference between the things that do happen and those that do not happen is a different difference from that between the things that can happen and those that cannot happen, so the reasons why the things that do not happen do not happen are logically different from the reasons why the things that cannot happen cannot happen. Since, however, what cannot happen does not happen, though what does not happen need not be what cannot happen, the reasons why something cannot happen are in fact also reasons why it does not happen, though the reasons why something does not happen need not be reasons why it cannot happen. In other words, a necessary condition of 'can' is also a necessary condition of 'is'; but not *vice versa*. Thus, on the one hand, the reason why a particular man did not refrain from stealing may be that, being a kleptomaniac, he could not refrain. But, on the other hand, the reason why a golfer fails to sink an easy putt need not be that he has momentarily lost his ability—and hence could not have sunk it—but that something caused him not to sink it. It may be true that certain footprints could have been made either by a panther or a puma even though the absence of any panther proves that they were not made by a panther.

Hence, in so far as the existence of a prior cause is a reason why something does happen and its opposite does not happen, it is not necessarily a reason why so-and-so can happen or why its opposite cannot happen. Not all causes compel.[60] When it is true to say that something could have happened but did not, it is reasonable to ask why it did not and, perhaps, to answer by mentioning the absence of a causal condition. But the reason why it did not happen is, *ex hypothesi*, not a reason why it could not have happened.

In so far as the doctrine of universal causality—and perhaps one version of Determinism—only states that everything that does happen has a cause, it is not incompatible with the view that the existence of the cause of something's happening—and, therefore, of its opposite's not happening—does not preclude that the opposite of that something could have happened. The absence of the causal conditions of X—which preclude X's actually happening—does not preclude that X could have happened any more than the absence of X itself—which precludes X's actually happening—precludes that X could have happened. Conversely, to assert that X could have happened does not deny the existence of causal conditions why X did not happen any more than it denies that X did not happen. If the possibility of X's happening and the absence of the causal conditions for X's happenings were incompatible, the possibility of X's happening would not differ causally from X's happening. But there is a world of difference here.[61]

Nor does it matter whether the 'could have happened' whose truth is compatible with the absence of those conditions which would have brought about the happening, is unconditional, as in 'He could have got in by the bedroom window', or conditional, as in 'He could have got in by the bedroom window, if he had had a ladder', since the condition thus introduced cannot properly be such as to change the 'could have' into 'would have'.

By parity of reasoning, the existence of the cause of X's happening does not preclude the possibility of that cause's non-existence; and, similarly, for the cause of that cause *ad infinitum.*

In order, therefore, to prove that what did not happen could not have happened, it would be necessary to prove not merely that there existed a cause of what happened but that this cause was such as to make what actually happened necessarily happen. Undoubtedly some causes are of this kind; a blocked carburettor not only causes the car to do less than 60 m.p.h., but also makes it impossible for it to do more. But not all causes have this double effect; a half-opened throttle causes the car to do less than 60 m.p.h., but does not make it impossible for the car to do more than 60 m.p.h. The throttle opening governs only the actual speed of the car; the state of the carburettor governs its speed-capability. In order to prove that what did not happen could not have happened, it would be necessary to prove not merely a law that everything, actual and possible, has a cause, but that any cause makes its effect both actual and necessary. And, though this may be the basic belief of Determinists, I know of no evidence for its truth.[62] Hence, there is no need to invoke any idea of a 'contra-causal freedom'[63] in order to explain how what was not could have been. By the law of causality it follows that if so-and-so happened, then what happened had such-and-such antecedents. But it does not follow from this that nothing other than what did happen could have happened, but only that nothing other than what did happen would have happened. Naturally, if something other than what happened had happened, as contrasted with the fact that it could have happened, then, by the law of causality, different antecedents would have been present.

The plausibility of the supposition that a condition of something's not happening is also a condition of the impossibility of its happening is, perhaps, partly due to a confusion[64] between the correct view that if a condition is sufficient for something, then it is impossible to have both the condition and the absence of that for which it is a condition and the incorrect view that if a condition is sufficient for something, then it is also impossible to have both the condition and the possibility of the absence of that for which it is a condition. Such a confusion arises, however, from transfering the 'must' (necessity) or 'can't' (impossibility) of e.g., 'If X is so, Y must be so' or 'If X is so, not-Y can't be so' from the *connection* between the two related items to the *second* item itself. The mistake of arguing from 'It is physically impossible for X to have happened when the causal antecedent of X was absent' to 'If the causal antecedent of X was absent, then it's physically impossible for X to have happened' is parallel to the traditional mistake, made perhaps by Plato, Hume and the Logical Positivists, of arguing from 'It is logically impossible for p to be false when p is known to be true' to 'If p is known to be true, then it is logically impossible for p to be false', that is 'p is a necessary truth'. In symbols $\sim Mp(\text{not-}aX\ .\ X) > (\text{not-}aX > \sim MpX)$ is invalid for the same reason that $\sim M(\text{not-}X\ .\ X) \rightarrow (\text{not-}X \rightarrow \sim MX)$ is invalid—where Mp=physically possible and aX = antecedent of X. For each of these has a consequent which absurdly asserts, in its contrapositive form, either, as in the former, that the physical possibility of one thing implies the actual existence of its causal antecedent, that is $MpX \rightarrow aX$, or, as in the latter, that the logical possibility of one thing implies the actual existence of itself, that is $MX \rightarrow X$.

Similarly, to say that one thing 'determines'[65] another, e.g. that the height determines the width or the price, implies that it is not possible for anything other than the second to go with the first. It does not imply that nothing other than the second is possible. Thus, the fact that the speed of the car is determined by the size of the engine and the pressure on the throttle does not entail that the car could not have gone faster than it did.

The debate between believers in Determinism and believers in Free Will has not, of course, been based solely on an assumption about the meaning of 'could have done otherwise'. Often what has been in dispute is not the logical difference between the impossibility of a different connection between two events and the impossibility of an event different from the second event, but the scientific and common-sense question whether all events, including those involved in the behaviour of humans and of sub-atomic particles, are necessarily connected in the way that many events are. In other words, it is true, as Hume suggested, that 'it is almost impossible to engage either in science or action of any kind without acknowledging the doctrine of necessity, and this inference from motive to voluntary actions, from characters to conduct'. The solution of this dispute is not dependent on any analysis of 'could have done otherwise' nor, perhaps, even on any purely conceptual clarification.

In his well known discussion of the Free Will problem, G. E. Moore (*Ethics*, ch. 6) asserted that if everything has a cause, then nothing could have happened except what did happen, while at the same time it is beyond doubt, as believers in Free Will assert, that many things which did not happen could have happened. He argued, therefore, that either the belief in universal causality is mistaken or 'could' is ambiguous.

It follows from our previous discussion that Moore was here doubly wrong. First, in his supposition that 'could' is ambiguous. His only evidence for this is that when 'could have' is used to make the undoubtedly true statement that many things which did not happen could have happened, it means 'should, if I had chosen'. This analysis, however, is, as we have seen, not only very doubtful in itself, but is impossible to apply to inanimate cases of 'could have, but did not'. Moore partly realizes this in his suggestion that 'The ship could have steamed at 20 knots, but did not' means that she would have, if the men on board her had chosen. But there are plenty of examples of 'could have' applied to inanimate objects where no reference to human beings is at all necessary or plausible, e.g. 'A tree could have crashed on the car'.

The fact is that, although Moore explicitly, and rightly,[66] extended his remarks about the indubitability of 'could have, but did not' to cover every type of object—and so said, e.g., 'we continually, when considering two events, neither of which *did* happen, distinguish between them by saying that whereas the one *was* possible, though it didn't happen, the other was *im*possible' (p. 208)—yet, because he was primarily interested in the problem of free will, he confined his analysis of 'could have, but did not' to examples with a personal subject, e.g. 'I could have walked a mile in 20 minutes this morning, though I did not', for which, of course, 'I should, if I had chosen' makes sense, though not the same sense. It may be that what led Moore to think that 'could' is ambiguous was his feeling of a difference, which I have already noted, between 'One can't have X and not-Y' and 'If one has X, one can't have not-Y', the former of which would exemplify the permissible law of causality, while the latter would exemplify the impermissible view that the opposite of what happens is impossible. Unfortunately, his early view that 'could' can be elliptical for 'could, if' or 'would, if' led him to seek the difference elsewhere.

Moore's second mistake is his thesis that, if 'could have' is not ambiguous, a belief in universal causality is incompatible with the belief that many things could have happened but did not. This thesis rests on the assumption that if everything has a cause, then nothing ever could have happened except what did happen. This is an assumption which I have argued is a mistaken interpretation of the law of universal causality, which holds, not that the absence of a given event makes it impossible for another event to occur, but only that it makes it that the other event does not occur. A second source of Moore's confusion here may be the assimilation of 'could have' and 'might have'. The most that the statement that everything has a cause could imply is not that it is *impossible for* anything other than what happened *to* happen but that it is *not possible that* anything other than what happened *did* happen. In other words even a strong version of universal causality would rule out, not that something else *could* have happened, but that something else *might* have happened, as well, of course, as ruling out that something else *did* happen. For, on this view, in so far as what did happen was caused to happen the antecedent factors did not leave open the possibility of any alternative and, therefore, ruled out the suggestion that such an alternative might have happened. But to hold, whether rightly or wrongly, that determinism shows that it is impossible that something will be otherwise, that is, that it is not true that it might be otherwise and, therefore, that it is certain that it will not be, is not to be committed to holding that it shows that it is impossible for it to be otherwise, that is, that it is not true that it can be otherwise and, therefore, that it is necessary that it should not be otherwise.

(D) CAN, MAY, MUST AND OUGHT

Just as the relations between *could* and *would* and those between *could have* and *did not* are both puzzling in themselves and the source of deep-seated puzzles in traditional philosophical beliefs, so also are the relations between *can* and *must*, and *can* and *ought*. Such relations will, however, be more fruitfully discussed later when the notions of *may*, *must* and *ought* have themselves been examined.

NOTES

1. 'Can' does not have different senses in these different exemplifications, *pace* Nowell-Smith (1960), pp. 87 ff.; O'Connor (1960); Aune (1962–3), p. 405 and (1970); R. Taylor (1960), pp. 78–89; (1966), ch. 4; Thalberg (1969), pp. 182–204; Wittgenstein (1958), p. 114; von Wright (1951b), p. 28. Ayers (1968), ch. 6 confuses the difference between human abilities (and even freedom) and inanimate potentialities with alleged different kinds, though not senses, of 'can'.
2. Lebrun, p. 54, finds no example of 'can have'.
3. There is no justification for holding that 'could have' is always subjunctive, e.g. Ayers (1966), pp. 113–20. Contrast Austin (1961), p. 163; Ehrman; Joos, ch. 6; Palmer, chs. 6 and 7.
4. The examples quoted by Ayers (1966) are no exception, since they refer to a (unspecified) particular time during a period and not to the period itself, e.g. 'At any time (or while I was king) I could have Ved'.
5. E.g. Honoré, p. 469; O'Connor (1960), p. 17; Locke, p. 253.
6. E.g. Palmer, p. 118.
7. E.g. O'Connor (1960), p. 17.
8. E.g. Honoré, p. 464.
9. E.g. Thalberg (1969), p. 188; Locke, p. 253.
10. E.g. Austin (1961), p. 175; contrast Thalberg (1969), p. 187.
11. *Pace* Mayo (1968), pp. 275–6; cp. Gert and Martin; Pears (1973), pp. 127–34.
12. E.g. Ryle, p. 246; Nowell-Smith (1954), p. 274; Locke, p. 247; Ayers (1966), p. 119; cp. Ehrman, p. 30; Joos, p. 187 for 'might'; Twaddell, p. 7; contrast Austin (1961), pp. 168, 171; Gallop (1968), pp. 255–6.
13. Contrast Locke, p. 255.
14. Pears, pp. 96–101, seems to be making a similar point in his distinction between the unconditional possession of a conditional power and the conditional possession of an unconditional power.
15. Pears, p. 92 seems to reject this type of case on this ground.
16. E.g. (1961), pp. 159, 161–2.
17. Contrast Osborn, pp. 711–13.
18. Austin uses present tenses, 'I shall if I choose', etc., throughout, but his whole discussion shows that his argument ought to hold equally for the past. It may, of course, be that he would have held different theses about 'I would, if I had *chosen*' and 'I would, if I had *wanted*' e.g. p. 161.
19. Austin's reliance on the first person present of the verb is as misleading here as it is in his remarks on knowledge (1961), pp. 66–7, which have led many critics to suppose that he took a performative view of knowing. Nowell-Smith (1960), p. 86, claims, though without argument, that 'could have' is more plausibly related to 'if' than is 'can'.
20. E.g. Moore (1912), pp. 198–201. O'Connor's suggestion (1960), pp.

51 ff., that the 'if' of both 'could if' and 'would if' gives a *necessary* condition is based on (1) confusing 'if I had not chosen' and 'if I had chosen not' and on (2) assuming that 'I could if I chose' is ever a natural way of saying 'I could not, if I did not choose'.

21. Cp. Moore (1912), pp. 201 ff.; Ewing (1964), pp. 171 ff.
22. E.g. Baier (1963), p. 23; Honoré, p. 470, Whiteley (1963), p. 93; Matthews, p. 132; Ewing (1964), p. 166; Locke, p. 248; Pears, pp. 96, 110 ff.
23. E.g. *Analysis* Problem, pp. 125–32; Baier (1963), pp. 20–1; O'Connor (1960), p. 4; Chisholm (1964), pp. 20–1; Thalberg (1962); Ayers (1968), pp. 119–24; contrast Osborn, p. 715.
24. E.g. Moore (1912), p. 211.
25. E.g. Moore (1912), p. 211; Baier (1963), pp. 20 ff.
26. E.g. Nowell-Smith (1954), pp. 274–8; Ewing (1964), p. 171; Thalberg (1962); Aune (1967); Chisholm (1964), pp. 22 ff.; Honoré; Moore (1912), p. 212; Ayers (1968), esp. pp. 69 ff.; Ayer (1954), ch. 12.
27. Cp. Joos, Ehrman, Palmer.
28. Cp. Austin (1961); Nowell-Smith (1960), p. 87.
29. Austin (1961), p. 167, seems to suggest, wrongly, that 'could have' used of the inanimate is a vulgarism for 'might have'; cp. Raab, pp. 63–4; contrast Ewing (1964).
30. Cp. O'Connor (1960), p. 9.
31. Cp. Austin (1961), p. 164.
32. Cp. Ayers (1968), p. 123.
33. Cp. C. A. Campbell, p. 455.
34. Cp. Ewing (1964), p. 173.
35. E.g. Baier (1963), pp. 20 ff.; cp. Locke, p. 255.
36. E.g. Baier (1963), p. 24; cp. Ewing (1964), p. 160.
37. Cp. Ewing (1964), p. 160; contrast Baier (1963), p. 25, whose view rests on a narrow interpretation of 'could if I had chosen' as confined to results due to skill.
38. E.g. Ewing (1964), p. 160.
39. E.g. Baier (1963), pp. 20–5; Locke says they are 'often virtually interchangeable', p. 255.
40. E.g. Ewing (1964), p. 160.
41. E.g. Baier (1963), p. 23.
42. Pears, pp. 110–16 seems to argue that it *is* more plausible to hold that 'I can if I choose' implies 'I shall if I choose' than that 'I can' itself implies 'I shall if I choose' precisely because 'I choose' does eliminate the last possible obstacle.
43. E.g. Moore (1912), pp. 211 ff.; Nowell-Smith (1954), pp. 274–8 and (1960); Honoré; Ewing (1964), pp. 170 ff.; Thalberg (1962); R. Taylor (1960); Ayers (1968), pp. 69 ff., contrast pp. 125–35.
44. E.g. Ayers (1968), ch. 2.
45. Cp. R. Taylor (1966), ch. 4; Ayers (1968), ch. 2; but contrast pp. 169–79.
46. Though Ayers (1968) accepts the analysis for what he calls 'natural

possibility and rejects it for what he calls 'possible for choice', cp. pp. 69 ff. and 125 ff.
47. Cp. Locke; Dore (1966), p. 138.
48. Cp. Lehrer (1961).
49. Cp. Lehrer (1960).
50. Cp. Whiteley (1963), p. 92.
51. E.g. Chisholm (1964), pp. 23–4.
52. E.g. Nowell-Smith (1954) and (1960). Nowell-Smith's versions seem rather confused; e.g. (1954), p. 291 he says ' "I could not have kept my promise because I was kidnapped" . . . translated into the hypothetical form becomes "I would have kept my promise if I had not been kidnapped" '. But this is clearly fallacious. 'I could not have killed him because I was out of the country' does not mean or imply 'I would have killed him if I had not been out of the country'.
53. E.g. Moore (1912), pp. 212–7; Nowell-Smith (1960); R. Taylor (1960); Honoré; Kaufman (1963). Contrast Kaufman (1962), who holds only that 'could' implies 'would, if'.
54. E.g. Aune (1967); cp. Dore (1962); contrast Scarrow; Kaufman (1963).
55. E.g. Pap and Scriven; cp. Austin's aside in (1961), p. 166, note.
56. E.g. Nowell-Smith (1960); Honoré; Aune (1967).
57. E.g. Chisholm (1964), p. 24.
58. Ewing (1964), pp. 163, 173.
59. Cp. Chisholm (1964), pp. 24–5, and (1967); Dore (1962); Lehrer (1966) and (1968). Ayers (1968), pp. 87–9 admits the difficulty. Aune's answer (1967) seems simply to beg the question—cp. Hunter—and his criticism (1970) depends on the alleged existence of different senses of 'could'.
60. Cp. Ayer (1954).
61. Cp. Aune (1962/3), pp. 408-10.
62. Cp. Ayers (1968), ch. 1.
63. E.g. C. A. Campbell, pp. 441–65.
64. E.g. R. Taylor (1960); contrast Aune (1962/3), pp. 408–13.
65. On 'determine', see Gallop (1962).
66. Ewing (1964), pp. 162–5 rightly stresses that the analysis of 'could have' should cover the inanimate as well as the animate.

Chapter Three

May

(A) MAY

Grammatically, 'may' goes with any verb, except some other modals, to express the idea that it is possible that something or somebody Vs or will V. 'May have' means 'It is possible that somebody or something did V'. For example, 'A may be (have been) at the barber's', means 'It is possible that A is (was) at the barber's; 'You may say that I should have acted differently' means 'It is possible that you will say that I should have acted differently'. The question[1]* 'May we be doing (have done) him an injustice?' means 'Is it possible that we are doing (did) him an injustice'; while the prayer 'May we do him no wrong' means 'Let it be possible that we do him no wrong'—compare 'May God have mercy on his soul'.

The past form 'might', meaning 'It is possible that somebody or something would V'—and the corresponding perfect form 'might have', meaning 'It is possible that something or somebody would have Ved'—sometimes signifies remoteness in time. This occurs in the habitual past—e.g. 'While we were there, we might sometimes play bridge after dinner' means 'While we were there, it is possible that we would sometimes play bridge after dinner'—and in reported speech—e.g. 'He said that somebody might have stolen it', where his words were 'Somebody may have stolen it'. Here 'might' is unconditional. More commonly, however, 'might' (or 'might have') signifies remoteness in reality, whether this is expressed conditionally as in 'If he had been younger, he might have succeeded' or tentatively as in 'There might be some left' or in 'Might it be in the drawer?'. The common use of 'might' instead of 'may' in questions could be either a sign of tentative-

* References to this chapter begin on p. 57.

ness or a means of avoiding confusion with a request. Thus we say 'May he take it?' for a request, but 'Might he take it?' either for a more tentative request or for a question. Further, just as 'You'll come' could be a prediction, an order, a question or an invitation, etc., so 'You might drop in on your way home' or 'The two parties might meet secretly' could be either a prediction or an invitation (or suggestion).

'May' is frequently used in asking for or giving permission for something, e.g. 'May I go?', 'You may'. When so used, it has certain grammatical peculiarities. For instance, the past forms 'may have' and 'might have' do not occur; and 'might' is used only in the asking for and the reporting of, but not in the giving of, permission. In giving permission, 'may' has also certain logical peculiarities. For instance, 'He may not speak English', used as the refusal of permission, negates the possibility of his speaking English, whereas, used as a conjecture, it suggests the possibility of his not speaking English. 'May' can be qualified in a statement of possibility in ways not allowable in the granting of permission; e.g. 'He may possibly, probably, easily, do it'. 'He may, but he won't' is queer as a statement of possibility, but proper as part of a permission.

But neither the grammatical[2] nor the logical[3] differences provide a reason for supposing a different sense or use of 'may' any more than similar differences provided reasons for a different sense of 'can'. They are to be explained wholly in terms of the different sorts of conditions which allow that something is possible. To ask for or to grant permission by the use of 'may' is to ask whether, or to state that, someone's wishes allow the possibility that something will be so, just as to ask for or grant permission by the use of 'can' is to ask whether, or to state that, someone's wishes allow that it is possible for something to be so.

A common misinterpretation of the meaning of 'may' and 'might' assigns them the sense of 'possible that' followed by the subjunctive rather than of 'possible that' followed by the indicative, that is, of 'possible that it should' rather than of 'possible that it is'. Thus, for example, G. E. Moore usually supposed[4] that 'X might V' is equivalent to 'It is possible that X should V'. Because he also usually took 'logically possible' to mean 'logically possible that it should', he usually linked 'might' with logical possibility. In these moves he is followed by I. Hacking.[5] When discussing 'causal possibility', on the other hand, Moore sometimes linked it with 'possible to V' and 'can V'. At other times, however, he linked it with 'possible that it should V', which, in agreement with his previous analysis, he usually—incorrectly in my opinion—expressed as 'might V'; but sometimes he—correctly in my opinion—expressed as 'can V'.

M. R. Ayers[6] is inclined to restrict 'possible for X to V' to 'possibility for choice' and to call 'possible that it should' 'natural possibility'. But there is no difference in possibility between 'It is not possible for him to do X' and 'It is not possible for him to have been killed, to know Z, to foretell p, to survive, etc.' though only the former entails choice.

The possibility expressed by 'may' and 'might' has the same relation to the actual as has that expressed by 'can'. Thus, to say that X may or might V, may have or might have Ved, does not imply—any more than X can or could V, can have or could have Ved—that X did, does or will V or that X did not, does not or will not V.

The falsity of 'X may have Ved' implies the falsity of 'X Ved'; and, therefore, the truth of 'X Ved' implies the truth of 'X may have Ved'. The truth of 'X may have Ved', however, leaves it quite open whether X did or did not V.

It is one common philosophical mistake[7] to suppose that 'X might have Ved'—or 'X could have Ved'—implies either that X did not V or that the speaker supposes that X did not V. If, wondering how someone got into my office, I suggest that a cleaner might have left a door unlocked, I am far from implying that she did not or that I suppose that she did not.

It is an opposite philosophical mistake[8] to suppose that 'X may have Ved' is incompatible with the falsity of 'X Ved'. Since the exposure of this mistake would be exactly parallel to the *reductio ad absurdum* disproof already given (pp. 7–8) of the thesis that 'X did not V' implies 'It is not possible that X Ved', it need not be repeated here.

One possible reason for the mistaken supposition that 'X may have Ved' is incompatible with the falsity of 'X did V' is a confusion of what it is *true* to say with what it is *proper* to say. Whether we know that X Ved or that he did not V or that he would have Ved if . . ., we can sometimes state definitely and properly that he could have Ved or that he could have Ved if . . . Whereas, if we either know or believe that A Ved or that A did not V, we cannot properly say that he may have Ved or that he might have Ved;[9] though we can say that he might have Ved if . . . In other words, to assert 'A may have Ved' and 'A might have Ved', unlike asserting 'A could have Ved' or asserting 'A could have, may have or might have, Ved if . . .', implies that the assertor cannot properly claim either to know or believe that A did V or that A did not V. 'A may have Ved, but he did not' is a queer thing to say, whereas 'A could have Ved, but he did not' is legitimate. 'A might have Ved, but he did not' is only sensible when the 'might' expresses remoteness in reality. *A fortiori*, 'It is improbable but true' and 'It is probable but false' are not permissible things to say.[10] This is not, however, to take the view[11] that to assert 'A may have Ved' or 'A might have Ved' implies or suggests that the assertor does know or think that A did not V nor, of course, that he does know or think that A did V. To say that 'p implies I do not know q' is not to say that 'p implies I know not-q'. The inappropriateness of saying these things, as contrasted with their truth, stems from the fact that the notion of *possibility* expressed by 'may' and 'probable' is, as I have mentioned, that of the possible existence of an actuality, whereas the possibility expressed by 'can' is the actual existence of a possibility. Although the actuality of anything is, as I have argued, just as compatible with the possibility of the actuality of its opposite as it is with the actuality of the possibility of its opposite, it would be queer for someone who was expressing his belief in the actuality of something to express at the same time his belief in the possibility of the actuality of its opposite in a way that it would not be queer to express at the same time his belief in the actuality of the possibility of its opposite. 'A may have Ved, but he did not' (or 'A did not V, but he may have') is, in this respect, analogous to 'I believe that A Ved, but he did not', which also, though a queer thing to say, is perfectly capable of being quite true.

Another reason for the mistaken supposition that 'X may have Ved' is incompatible with the falsity of 'X did V' is, perhaps, a confusion—analogous to that which we saw to underlie one of the arguments for supposing that nothing but what did V could have Ved—between the truism that if X did not V, then X did not V and the mistaken supposition that if X did not V, then it's not the case that X may have Ved. Such a confusion could arise if both the truism and the mistaken supposition were expressed carelessly as 'If X did not V, then it is not possible that X did V' where 'possible' is being used in the former to connect the two clauses of a proposition and in the latter to express the second clause of a proposition.

Most philosophers[12] who have noticed that 'may' (or 'possible that' and the indicative) and 'can' (or 'possible to') are different have suggested that 'X may have Ved' (or 'It is possible that X Ved') means that it is not known that X did not V. Hence, 'may' is said to express 'epistemic' possibility. Such an analysis, however, has several difficulties. First, what is not known at one time may become known at a later date. But we have seen that it makes no sense to say 'It once was possible that X Ved, but that is no longer possible' as contrasted with saying 'It once was possible for X to V, but that is no longer possible'. Secondly, those versions of the thesis that interpret 'may' more narrowly in terms of what is not known to the speaker fall into Moore's mistake, criticized earlier, of supposing that, since what is not known to me may be known to you, the same thing can be both possible and not possible. Thirdly, the thesis confuses the fact that 'X may have Ved' is appropriately used only by someone who does not know that X did not V or, indeed, who does not know that X did V with the false supposition that 'X may have Ved' is true only in these circumstances. Fourthly, the thesis confuses the evidence for supposing that it is possible that X Ved with the circumstances that make it possible that X Ved. We shall, in our later discussion, see this confusion writ large in subjective theories of probability. Finally, the theory owes much of its plausibility to certain assumptions about the relation between *knowledge* and *certainty*, which, I shall argue, are without foundation.

(B) CAN AND MAY

The relation of 'can' to 'may' is exactly the same as that of 'possible to' to 'possible that' with the indicative, namely the relation of existential to problematic possibility.[13] Hence, most of its characteristics have already been brought out in our earlier remarks on the latter. Since, however, numerous erroneous opinions have been held, both by philosophers[14] and by linguists,[15] on 'can' and 'may', it is worth underlining a few of these differences as they are expressed in these forms. What may have been so must be such as could have been so and might have been so; but of neither what could have been so nor of what might have been so is it necessarily true that it may have been so. Hence, sceptical arguments designed either to show that it could have been that X did not V or, what is different, to show that it might have been that X did not V, go no way to show that it may have been that X did not V and, therefore, go no way to show that X did not V or that it is not certain that X did V.[16] Similarly, arguments to show that it is not necessary for X to V do not show that it is not certain that X Ved.

Analogously, what might have Ved must be such that it could have Ved, but it is not necessary that what could have Ved might have Ved. Hence, if it is false that X could have Ved, it is false that X might have Ved, but it can be false that X might have Ved without being false that X could have Ved.[17]

We saw earlier (pp. 15–16) that G. E. Moore rightly objected that Russell's scepticism rested on the confusion between its being possible for X to V and its being possible that X Ved. His objection also contains the allegation that Russell had confused two different uses of 'may'. This allegation, however, rests on Moore's mistaken belief that, e.g., 'It is possible for a human being to be of the female sex' can be expressed as 'Human beings *may* be of the female sex'. Hence, he was led to suggest that 'may' is being 'used in a totally different sense in "Human beings may be of the female sex" and in "This human being may be of the female sex"; and, further, that the sentence 'Human beings may be of the female sex' does not mean '*Every* human being *may* be', but only '*Some* human beings *are*'. But, in fact, 'It is possible for a human being to be of the female sex' is equivalent to 'Human beings *can* be of the female sex' and not to 'Human beings *may* be of the female sex'. Furthermore, 'Human beings may be'—which is equivalent to 'It is possible that human beings are'—does mean '*Every* (or any) human being *may* be' just as much as 'Human beings *can* be' means '*Every* (or any) human being *can* be'. For instance, 'Human beings, but not animals, *may* be born with an immortal soul'. Russell's fallacy[18] was in fact to move from 'It is possible for an X (or this X) to be Y' to 'It is possible that this X is Y'; or, in other words, to move from 'An X (or this X) *can* be Y' to 'This X *may* be Y'.

Having made certain that in the circumstances A *could* have Ved, I may then wonder whether A *would* have Ved (or whether he did V). My uncertainty whether he would have Ved, as contrasted with my certainty that he could have Ved, finds its natural expression in the conjecture that he might (or may) have Ved.

Having protested that I could have been getting on with my article instead of spending the afternoon at a committee meeting, I may be asked 'But would you have got on with it?' and be able only to reply feebly 'Well, I might have'. That is, it may be both true and known that something *could* have been so without its being either true or known that it *might* have been so, much less that it *would* have been so. This is probably why 'A could have Ved', as contrasted with 'A may (might) have Ved', sounds strange with mental verbs such as 'be interested', 'wish', 'like', where there can be no doubt that it is possible for A to V, though much doubt whether it is possible that A did or would V. While it makes sense to say 'He may (might) have been able to V' and 'It may (might) have been possible to V', it makes no sense to say 'He could have been able' or 'It could have been possible to'. 'X might (or may) have Ved', unlike 'X could have Ved', is a direct, though hesitant or tentative, answer to the question 'Did X V?' or 'Would X have Ved?'. It qualifies the answer in the way that 'maybe' and 'perhaps' do. This is why verbs of hoping, fearing, wondering, etc., which are often tentative, are also often followed by 'might' instead of 'would'. Just as 'X may have Ved' is a speculation that it did V, so 'X might have Ved' is a speculation that it would have Ved, e.g. 'It might have worked for a little while, but it would not have lasted'. There is sometimes a suggestion that what might have Ved (or might have Ved if) would have Ved (or would have Ved if), but no suggestion that what could have Ved would have Ved. Alternative tentative answers to the question 'What *would* they have done . . .?' take the form 'They might have done X or Y or Z' and not the form 'They *could* have done X or Y or Z', which would be an answer to 'What *could* they have done . . .?'. This is why, when it is known that X did not V, 'X might have Ved', unlike 'X could have Ved', can only be used hypothetically. On the one hand, 'I would have done it, if I could have' makes sense, but 'I would have done it, if I might have' does not. On the other, we sigh for what might have been, not for what could have been.

The evidence used to *prove* that so-and-so could have Ved, that is, that it was possible for it to have Ved, is unlike the evidence to *suggest the hypothesis* that so-and-so may (or might) have Ved, that is, that it is possible that it did V or would have Ved. We cannot properly argue from the fact that so-and-so is not *certain* because the facts *might* go against it to the (mistaken) assertion that 'it would still not be certain, because the facts *could* go against it'.[19] The fact that the suspect had the necessary skill and opportunity shows that he *could* have committed a crime; the fact that he had these together with an appropriate motive suggests that he *may* or *might* have done it. It is often this addition of 'if he had wanted' which enables us to move from the truth of 'could have' to the truth of 'may have' or 'might have'. To make it possible for something to V is to enable it to V or to make it liable to V, whereas to make it possible that it will V is to make it likely to V. To say 'If A had known that p, he could have done X' suggests that A's knowledge would have put him in a position to do X, whereas to say that 'If A had known that p, he might have done X' suggests that the knowledge would have given him a reason for doing X. In general, the protasis of a 'could have' apodosis commonly states the *conditions* which make the apodosis true, e.g. 'If the hole had been bigger, the water could have drained away', 'If he had been helped, he could have lifted it'; the protasis of a 'might have' apodosis, on the other hand, state the *grounds* on which the apodosis is advanced, e.g. 'If the hole had been bigger, the water might have drained away', 'If he had been helped, he might have lifted it'.

Hence the strangeness of 'If he had been asked, he could have done it' as contrasted with the propriety of 'If he had been asked, he might have done it'. His having been asked provides a likely ground for supposing that someone would do something; it is not normally an enabling condition for doing it. Conversely, the grounds for the suggestion that so-and-so *might not* have Ved may be replaced by a knowledge of conditions in which so-and-so *could not* have Ved.

Whereas the same conditions may make it possible for X to have happened as make it possible for X not to have happened, e.g. a type of spring makes it possible for the door to have been left open or shut; the evidence that it is possible that X did happen or would have happened is some evidence against the possibility that X did not happen or would not have happened, e.g. the lack of draught which shows it is possible that the door was or would have been shut counts against the possibility that the door was not or would not have been shut.

For similar reasons, the contradictory of 'X could have happened' is 'X could not have happened' and of 'It was possible for X to happen' is 'It was not possible (or it was impossible) for X to happen'. The assertion that the possibility did not exist is the negation of the assertion that the possibility did exist. The contradictory of 'X may (or might) have happened', on the other hand, is not 'X may (or might) not have happened', which asserts that 'It is possible that X did not happen (or would not have happened)', but 'It is not true that X may (or might) have happened', which asserts that it is not possible that X did happen or would have happened. As the grammarians say, 'not' negates the auxiliary with 'can', but the main verb with 'may/might', except for the 'may' of permission.[20] This is why 'scarcely' and 'hardly' can qualify *could have*, but not *may have* or *might have*. 'X could not have happened' expresses unreservedly the impossibility of X's having happened, whereas 'X may (or might) have not happened' expresses hesitantly the possibility of X's not having happened.

NOTES

1. Palmer, p. 119, wrongly denies that 'may' can be used in questions.
2. *Pace* Palmer, pp. 118, 130.
3. *Pace* some deontic logicians.
4. E.g. (1962), pp. 184–8; cp. (1959), pp. 223, 234, 250.
5. (1967); cp. Lebrun, pp. 44–60, 67–71.
6. (1968), pp. 12–13.
7. E.g. Ryle, p. 246; Nowell-Smith (1954), p. 274; Locke, p. 247; Ayers (1966), p. 119; cp. Ehrman, p. 30, Joos, p. 187 for 'might'; Twaddell, p. 7; contrast Austin (1961), pp. 168, 171; Gallop (1967), pp. 255–6.
8. Cp. R. Taylor (1957), pp. 11, 20.
9. Cp. Moore (1962), pp. 184–8 and (1959), pp. 232–5, on 'may'. Moore, however, allows the unconditional use of 'A might have Ved' even when it is known that A did not V; e.g. (1962), p. 185; (1959), p. 230.
10. Cp. Toulmin, pp. 54–7; contrast Kneale (1949), pp. 9–10.
11. As those referred to in footnote 7 do.
12. E.g. Moore, Hacking, Ayers, etc.
13. Cp. Ehrman, chs. 1 and 2; Joos, ch. 6.
14. E.g. Malcolm (1963), pp. 28–9 and (1962), pp. 29–31. Moore (1962), pp. 184–6, links 'might' to logically possible and to 'possible that it should', 'may' to epistemic possibility and 'could' to causal possibility; cp. (1959), pp. 223, 234, 250, where he seems to equate 'might have' with 'it would be possible to' and with, what he takes to be its equivalent, 'it is possible that it should'. In (1962), pp. 187–8, however, he wavers between linking causal possibility to 'might have' and linking it to 'could have'. Hacking (1967), p. 146 makes 'might have been' a necessary condition for 'logically possible' because he equates 'logically possible' with 'logically possible that there should have been' instead of with 'logically possible for there to have been' and, like Moore, takes 'might' as equivalent to 'possible that it should'. Instead of my threefold division of 'could have' ('possible for it to be'), 'may have' ('possible that it did') and 'might have' ('possible that it would have'), Hacking has two unclearly related pairs of contrasts, might/may and possible for/possible that. Cp. Hampshire (1965), pp. 11 ff.; Fogelin, ch. II.
15. E.g. Lebrun.
16. Cp. Hacking (1967), p. 150. Contrast, on the one hand, Ryle, p. 131, who holds 'roughly, to say "can" is to say that is is not a certainty that something will not be the case'. But this means that 'It is possible for p to be true' implies 'It is not certain that not-p is true' and, hence, implies 'It is possible that p is true'. Ayers (1968), pp. 14–15, 27, 33, 39, on the other hand, denies both that 'possible to' implies 'possible that' and that 'possible that' implies 'possible to'. But on the latter denial 'X may (or might) be so, though it could not be so' would not be a contradiction. And this seems not merely

something which it is *improper to say* (cp. Ayers (1968), p. 39), but false and inconsistent with the truth that if it is necessary for p to be true, it is certain that p is true.

17. E.g. Anscombe (1956), p. 14.
18. Conversely, determinism sometimes moves from the impossibility that it will be otherwise—that is, the certainty that it will not be otherwise—to the impossibility for it to be otherwise, that is, that it cannot be otherwise, cp. Ayers (1968), pp. 26–7.
19. E.g. Anscombe (1956), p. 14.
20. Cp. Ehrman, p. 26; Lebrun, p. 47.

Chapter Four

Probability

For both the possibility that something should be so, that is, the possibility of 'can', and the possibility that something is so, that is, the possibility of 'may', circumstances may be such that *either* they exclude the possibility of something *or* they exclude the possibility of its opposite *or* they allow both its possibility and the possibility of its opposite. Where they exclude the possibility of something, they make it *impossible*—though the exclusion of the possibility of 'may' is usually expressed by 'not possible'. Where, however, they exclude the possibility of something's opposite they make the thing itself either *necessary* or *certain*, depending—as we mentioned earlier and shall see in detail later—on whether the possibility that is excluded is the possibility that it should be so ('can') or the possibility that it is so ('may'). *Probability* enters at this stage which is intermediate between the exclusion of the possibility of something and the exclusion of the possibility of its opposite, that is, the stage where there exist both the possibility of something and the possibility of its opposite.

To say that something is 'probable', or 'likely', is to indicate the degree to which circumstances favour its being so in comparison with the existence of its possible alternatives. Sometimes this can be done precisely, e.g. 'a probability of 1/6 or 60 per cent'; sometimes it can be done only roughly, e.g. 'pretty, very, highly probable'. The limits of such a range are, on the one hand, the complete exclusion of possible alternatives and, on the other, the complete exclusion of the thing itself. Probability theorists usually refer to these as 'a probability of 1' and 'a probability of 0' respectively; but this is misleading, since these, like certainty and impossibility, are not the highest and lowest degrees of probability, but its limits.

There can be degrees of probability, but not of possibility.

Something can be highly probable or extremely improbable, but not highly possible or extremely impossible. One thing can be more or less probable, but not more or less possible than another. The probability, but not the possibility, of something can increase or decrease. Its possibility can only appear or vanish. We can ask *how* probable, but not *how* possible, something is. The undoubted propriety of 'This is no more possible than that' is only an apparent exception analogous to 'This is no more right (correct) than that' or 'He is no more the King of Siam than I am'. Further, 'slight', 'faint', 'vague' refer to the indistinctness, not the degree, of possibility that something has.

The relation of probability to possibility is parallel with that of confidence to belief. Just as something cannot be probable unless it is possible, so one cannot be confident that p unless one believes that p. Conversely, just as it always makes sense to ask how probable is something which is possible, even if the answer is 'most improbable', so it always makes sense to ask how confident one is about something which one believes, even if the answer is 'not at all confident'. Just as there are degrees of probability, but not of possibility, so one can have degrees of confidence, but not of belief.

The possibility which is relevant to probability—and, hence, to certainty—is the possibility that something *is* so, that is, the possibility expressed by 'may', and not the possibility that something *should be* so (or the possibility for something to be so), that is, the possibility expressed by 'can'.

Grammar and idiom show this clearly.[1]* We speak of its being 'probable or certain that it is', but not of its being 'probable or certain for it to be' or 'probable or certain that it should be'. 'He (It) is certain to be' is only an apparent exception. For it is not analogous to the sensible 'It is possible for him (it) to be', nor does there exist a sensible co-ordinate of the form 'He (It) is possible or probable to be'. 'He (It) is certain to be' is, in fact, equivalent to 'It is certain that he (it) is or will be' and is co-ordinate with 'It is possible that he (it) is or will be'. 'He (It) is certain to be' is exactly analogous to 'He (It) is likely to be', which is equivalent to 'It is likely that he (it) is or will be'.

Conversely, 'possible' in 'as much (quickly, clearly, closely)

* References to this chapter begin on p. 73.

as (is) possible' alternates with 'necessary', but not with 'probable' and 'certain'. We speak of its being 'necessary for something to be' or 'necessary that it should be', but not of its being 'necessary that it is'.

We saw earlier (pp. 12–14) that the adverb 'possibly' seems to be almost invariably an indicator of the possibility of 'may'. Here it is an idiomatic substitute for 'possible that' + the indicative, as in 'He possibly went by train' instead of 'It is possible that he went by train'. 'Perhaps', 'probably' and 'certainly' are always possible alternatives to 'possibly', whereas 'necessarily' only rarely is in the affirmative form.[2] In the negative form, the difference is clearer, since 'possibly', like 'probably' and 'certainly', almost always precedes the negative, as it does the affirmative, whereas 'necessarily' usually follows the negative, as in 'He possibly (probably, certainly) did not believe what you said' contrasted with 'He did not necessarily believe what you said'.

Incidentally, one source of the fashionable Propensity Theory of Probability may be a confusion of the correct view that for it to be probable that something is so is for there to be some possibility of the 'may' sort, with the incorrect view that it is for there to be some possibility of the 'can' sort. The further assimilation of 'can' to 'power' or 'potentiality' makes it sound plausible that for it to be probable that something is so is for there to be a propensity for it to be so. It cannot be probable that X is Y unless it is also possible that X is Y and, *a fortiori*, possible for X to be Y; nor can it be possible that X is Y, though it can be possible for X to be Y, unless there is some answer to the question how probable is it that X is Y, even if the answer is 'most improbable'.

We shall see later the importance for understanding the notion of probability of remembering that, because the possibility of 'can' expresses the present, past or future possibility of something's being so, while the possibility of 'may' expresses the possibility that something is, was or will be, so, tense differences in the former can qualify the possibility itself, but in the latter only what it is a possibility of. Thus, as we saw, we can say that it is, was or will be, possible for something to be, and that it is possible that something was or will be; but not that it was or will be possible that it is. Hence, also, 'became' can be used with 'possible to V', but not with 'possible that it Vs', because 'became' is intermediate between 'was' and 'is'. What it was possible to do in the tenth century, e.g. to travel without a passport, may have become impossible in the twentieth century; but it makes no sense to say that it was possible in the tenth century that the Ptolemaic system was correct, but by the twentieth century this became no longer possible.

The notion of *probability*, like that of *possibility*, is expressed in English in various forms, nominal, adjectival and adverbial, of which the adjectival, as it occurs in the impersonal 'It is probable', can be taken as basic. Thus, to say that that probably did happen or that this probably will happen is to say that it is probable that that did happen or probable that this will happen.[3] To say that so-and-so is probable or that this is the probable such-and-such is to say that it is probable that this is or will be so-and-so or probable that this is or will be the such-and-such. If there is a probability of so-and-so or that such-and-such, then it is probable that so-and-so or such-and-such is or will be.

What we characterize as possible, probable or certain are events, happenings, occurrences, states of affairs, objects and facts. Death, ruin, conviction, success, the defeat of one's opponents, the second coming of Christ, that inflation will continue, that Mars is uninhabited, that Bacon did not write *Hamlet*, or that I have misunderstood Bernoulli's Theorem can all be possible, probable or certain. Although it can be probable that a proposition is true, a proposition itself cannot, despite what most philosophers say,[4] be probable. The common philosophical talk about 'probable propositions' is based on several errors. First, there is an assimilation of the view that the truth of a proposition can be probable to the quite different view that a proposition can be probable, just as there is an assimilation of the view that the truth of a proposition can be believed or known to the quite different view that a proposition can be believed or known. Secondly, there is the assumption[5] that what is probable when it is probable that p is the proposition that p; but this is no more correct than the assumption that what is regrettable, intolerable, hoped or feared when it is regrettable, intolerable, hoped or feared that p is the proposition that p. Thirdly, 'a statement of what is probable' is confused with 'the probability of a statement', so that what states a probability is itself supposed to have a probability. But what states a probability or states what is probable need no more be probable than he who drives fat oxen need himself be fat. Fourthly, 'probable proposition' is mistakenly modelled on 'probable explanation, cause or result'. For, whereas, 'a probable explanation, cause or result' is what is probably an explanation, cause or result, 'a probable proposition' is not what is probably a proposition. Finally, talk about the probability of a proposition may be a survival of the medieval usage which adhered to the etymological derivation of 'probable' from 'probabilis', meaning 'approvable or plausible'. It is only in this obsolete medieval sense that a proposition could be probable or that the word 'probable' could have an evaluative role.[6]

The belief that what is probable is a proposition may be a main source of the attempt to co-ordinate probability with truth, and, hence, of the suggestion[7] that 'the phrases "It is true that" and "It is probable that" are in the same line of business' and of the quarrel[8] whether something can be 'improbable but true'. No such co-ordination between probability and truth exists. What can be true or false is a proposition; what can be probable or improbable, like what can be possible or certain, likely or about to happen, is not. If it is true or plausible that p, it is true or plausible to say that p. But if it is probable that p, it is not probable to say that p any more than if it is likely, or if there is a chance, that p, then it is likely, or there is a chance, to say that p. If it has been truly or plausibly said that p, it follows that what has been said is true or plausible; but if it has probably been said that p, it does not follow that what has been said is probable. The limits of the probable are not the true and the false, but the certain and the impossible. 'Probability' does not signify[9] 'degrees of partial truth'—whatever that may mean.

Something's being possible or probable is not incompatible with its not being so, nor is its being improbable or possibly not so—as contrasted with its being not possibly so—incompatible with its being so. There is nothing strange, much less impossible, in the occurrence of the improbable.[10] After all, each hand of cards dealt in a game consists of an astronomically improbable sequence —which we don't comment on unless it is significant—but is nevertheless one which occurs. Likewise, 'It is probable that Tennyson was the author of that poem' no more implies that

Tennyson was the author than 'That poem was probably written by Tennyson' implies that it was written by Tennyson or than 'The probable author of that poem was Tennyson' implies that the author was Tennyson.

To insist that something's being probable is incompatible with its not being so would be to claim that whatever is probably so necessarily is so and that what is not so can have no probability of being so. This mistake is strictly analogous to that of supposing that what is not so can have no possibility of being so. That this latter is a mistake was shown earlier (pp. 7–8) by a *reductio ad absurdum* argument which I need not repeat here.

The belief [11] that what is probable is incompatible with what is not so may be due to confusing the undoubted inappropriateness of saying 'This is probably so, although it is not so' with the supposed falsity of what is said. The fact that no one who thought —much less someone who knew—that Tennyson was not the author of that poem would speculate on the possibility either of his being or his not being the author does not preclude the possibility or the probability of either of these. We are often perfectly happy to say that although something could be so, it is most unlikely to be, most improbable that it is, so.

We must distinguish[12] between the probability of something and an estimate of such a probability. An estimate of probability is not a probability any more than an estimate of length is a length. An estimate of the probability of X is made relatively to the evidence at one's disposal or on the basis of what one believes to be evidence, but the probability of X is relative to the factors for and against X, whether these are known or not.[13] Not being in a position to say at a particular time whether so-and-so is probable or certain does not entail that so-and-so is not at that time probable or certain. What makes one say, believe or estimate that so-and-so is probable, that is, one's grounds or reasons for saying this, is one's evidence: but what makes so-and-so probable are various features, facts, circumstances, etc. If a recently discovered letter makes it probable that Tennyson wrote the anonymous poem, it is the letter and not its being discovered that makes Tennyson's authorship probable. One can talk sensibly of the probability of so-and-so in such-and-such circumstances or of an estimate, made on such-and-such evidence, of the probability of so-and-so. But it is a philosopher's solecism to talk either of the probability or of an estimate of the probability of so-and-so on such-and-such evidence. Another solecism with the same roots is to speak of some things being 'probable for me but not probable for him' on the ground that the evidence at my disposal is different from that at his. What is meant is that what seems probable to me need not seem so to him. A change in one's evidence, that is in the evidence available to one, may lead to a change in one's estimate, so that what once seemed probable no longer seems so and is no longer asserted to be so. But only a change in the actual factors for or against X can lead to a change in the probability of X.[14] A revision of one's estimate of probability is not a change in the probability. It may once have seemed probable that there is life on the moon, but we now know that this is most improbable; but the probability of there being life there has not changed. The probability, on the other hand, that men will land on Mars before the end of this century is now not merely something that seems greater in the light of recent lunar successes; it is something which is greater. During a game of cricket the efforts of one team may make it possible, then probable and finally certain that their opponents will be all out before lunch.

What makes it possible or probable that something is so is not a subject for philosophical investigation. It is something which depends on the circumstances of each case. In mathematical probabilities there is no difficulty because we are given by definition all the positive and negative possibilities and the factors in favour of each, e.g. the probability of a perfect die turning up on any one face is the ratio between that face and the total number of its faces. In science and everyday life we have to discover, and so can be mistaken about, the possibilities and their favouring factors, e.g. whether the numbers of previous male and female births are the only factors relevant to the probability of my child's being a boy or girl, whether the mortality tables for men, for philosophers, for Irish-Canadians, are the only factors relevant to the probability of my living until I am 80. In mathematics we can mis-estimate only because our calculations or principles are wrong; in science even our data may have been wrongly selected.

What it is of which the probability is at issue relates to its specification,[15] e.g. as a randomly shuffled pack or as a pack with the court cards at the bottom, as a poem found amongst Tennyson's papers or as a poem with such and such literary characteristics. The probability of my living to be 80 will normally relate to me as a sedentary academic, the son of long-lived parents, hard working but of regular habits, a non-smoker but partial to fast driving. It will not normally relate to me as someone who will have a road accident next year or whose name is on the Mafia's current murder list. The probability of its being dry and sunny tomorrow relates to tomorrow as a day immediately following one with such-and-such weather characteristics or as an English June day or as both.

The common philosophical view that the probability of so-and-so's being such-and-such is relative to the evidence for this is due to confusing this relation with the relation between the probability of so-and-so's being such-and-such and the specification of so-and-so. In so far as it is known or believed that I am a non-smoker, the probability of my living to be 80 will properly be *estimated* to be so-and-so, but it is in as much as I am a non-smoker that the probability *is* so-and-so. In a similar way the question whether what I did was intentional or not is a function of the description of what I did. I may have intentionally set alight the package in the grate, which unknown to me contained bank-notes, but I did not intentionally set alight the bank-notes.

On the other hand, to say that that whose probability is at issue relates to its specification is not to say that what is at issue is the probability of some proposition containing this specification. To say, rightly, that 'under one description a thing may be said to have a probability of possessing some characteristic, whereas under another it would be plainly false to say that it had' is, *ex hypothesi*, not to say, wrongly, that 'probability does not attach to things, only to propositions and propositional functions'.[16] To say this latter would be like saying that because my action under one description, e.g. playing the piano, can be explicable in terms of its purpose or be absent-minded, whereas under another description, e.g. practising for a concert, it can be neither of these, therefore, it is not really actions but descriptions of actions that are absent-minded or explicable by their purpose.

Our idioms make clear that there can be no change in probability where only the available evidence and not the circumstances change, e.g. about the past or present. Thus, it makes no sense[17] to say 'In 1900 it *was* probable that Tennyson was the author of that poem, but by 1960 that *became* improbable'. What we do say is 'In 1900 it *seemed* probable that Tennyson was the author, but that is now thought (or seen) to be improbable'. 'Probable' is, in this respect, like 'possible', 'likely', 'definite', 'unfortunate', 'surprising', 'strange', etc., and unlike 'manifest', 'alleged', 'obvious', 'known', etc. It may be unfortunate, surprising or likely, just as it may be possible or improbable, that Caesar never invaded Ireland; but it makes no more sense to ask whether in 1885 it was—or whether in 1985 it will be—possible or improbable that he never did than to ask whether in 1885 it was—or whether in 1985 it will be—unfortunate, surprising or likely that he never did; though we can properly ask whether in 1885 it was—or whether in 1985 it will be—manifest, alleged or known that he never invaded Ireland and we can also properly ask whether in 1885 it seemed—or in 1985 it will seem—unfortunate, surprising or likely that he never did.

Where, on the other hand, the circumstances, as well as the evidence, can change, so can the probability. Thus we can assess the changing probability of what is in the future. Though even this may be expressed in terms of the probability, chances or likelihood of so-and-so's occurring rather than by different tenses of the impersonal 'It is probable'. Thus, we can say that our chances of winning the match were greater, or that it was more likely or that it became more probable that we would win, after the collapse of our opponents' opening batsmen than before. But even here it seems less natural to say 'At the beginning of the match it *was* probable that we would lose, but later on it *became* probable that we would win'.

All this is borne out by the ways in which we speak of 'chance' and 'likelihood'. The chances of such-and-such happening are typically something that can change, increase, decrease, improve, worsen, remain or vanish. It is significant in this context that we speak commonly of the chances of what has not yet happened and for which, therefore, there can yet be a change in the circumstances, but not usually of the chances of what has already happened and for the occurrence of which there could only be a change in the evidence. Furthermore, new circumstances may make something more or less likely to happen, but new evidence can only make something *seem* more or less likely either to happen or to have happened. Estimates of the probability or likelihood both of what has been and of what will be are tensed because such estimates can be past, present or future; but while there can be a past, present or future probability or likelihood of what is yet to be, there can be only a present probability or likelihood of what has been.

The common philosophical confusion of probability with estimates of probability which has led to the mistaken supposition of two senses of 'probable',[18] to a subjective analysis of *probable* in terms of 'rational assent'[19] and to an assimilation of the theory of probability and what is often called 'confirmation theory',[20] is partly due to the fact that to *say* that X is probable or that the probability of X is so-and-so is often to *express* one's estimate of its probability, just as to say that the length of X is so-and-so is often to express one's estimate of its length. Hence, philosophers[21] often confuse an interest in the conditions which are appropriate for *saying* that something is probable with an interest in the conditions in which something *is* probable just as they confuse the conditions for the appropriateness of saying it is possible that p—e.g. that they do not or cannot know that not-p—with the conditions for the possibility that p. But what one states when one says so-and-so is not what one expresses when one says it and what one is stating when one says that the probability of X is so-and-so is the probability of X just as what one states when one says that the length of Y is so-and-so is the length of Y. Stating that so-and-so is probable should not be assimilated to any purpose for which stating it is commonly or normally done, e.g. to give guarded guidance or to express one's partial belief as opposed to one's uncertainty. The purposes for which people say that something is probable are no more part of the meaning of the word 'probable' than the purposes for which people say that they know things are part of the meaning of the word 'know'.[22] Nor is stating the probability of X, which is what one does when one says 'The probability of X is so-and-so', or even stating that X is probable, which one does when one says 'X is probable' or 'probably X', to be confused[23] with stating guardedly or in a qualified way that X is so. People who wish to be cautious about what they say no doubt often say 'X is probably (or possibly) so' instead of 'X is so'. But this shows not that they are guardedly or with qualifications committing themselves to X's being so, but that they are unguardedly and without qualifications committing themselves to the probability of X's being so. 'X is probably so' is not a guarded or qualified version of 'X is so'; it is a guarded or qualified alternative to it. X's not being so does not refute the assertion of the probability of X's being so, whereas it would refute the assertion, however guarded, of X's being so. 'X is probably so' is no more a kind of assertion that X is so than is 'X is conceivably so'. Furthermore, such a view of probability, like 'illocutionary' theories of other concepts, is inapplicable to statements in the past tense or in conditional clauses. To say '30 years ago it was most improbable that men would ever land on the moon' is not to assert in a guarded way that men would never land on the moon.

Consider here a partial analogy between 'probably' and 'approximately'. 'There were probably 150 people in the hall' and 'There were approximately 150 people in the hall' would ordinarily be said by someone who did not know whether there were or were not 150 people there; yet neither is incompatible either with there being or with there not being 150 people in the hall. Hence, the insistence that either there were 150 or there were not is no more relevant to probability than to approximation. Both would be natural expressions of a cautious or guarded estimate of the size of the audience, yet both state something about the audience, the one its probable size and the other its approximate size. There is no more reason for denying that an audience can have a probable size as well as an actual size than that it can have an approximate size as well as an exact size. It does not, of course, have four sizes—which would mean four 'actual' sizes—any more than a room with an attractive size, an alleged size, an estimated size and an intended size has four sizes. Furthermore, just as some approximations, e.g. 'between 140 and 160' are better than others, e.g. 'between 100 and 200', so some estimates of probability are better than others. But it is just as mistaken to suppose that the best estimate of probability gives either certainty or actuality as it would be to suppose that the best approximation gives the actual size.

Another reason for the existence of the subjective view of probability may be that the word 'evidence' can, perhaps, be used either to mean the actual factors for or against something or only those factors which are offered as being, or are known to be, for or against it. To say that 'probability is relative to the evidence' would then be ambiguous as between the correct view that a change in the evidence (i.e. actual factors) can alter the probability of something and the incorrect view that such an alteration can be made by a change in the evidence (i.e. the factors offered or known). It is clear that subjectivists make probability relative to the known evidence. But what makes probable, e.g., Tennyson's authorship of a poem is that the style of the poem is his, that the date is that of his prolific period, that he did write on this topic, that there are references to such a poem in his letters, etc. and not merely those features which are known or believed in.

The evidence appealed-to may, for some philosophers, have usurped the actual evidence or factors in the role of that to which probability is relative because of their belief in a determinism according to which everything that actually is is certain and is made certain by what precedes it. For them 'probability' is only an expression of ignorance.[24] But even if such determinism does exist, it is not reflected in the current concept of probability. First, it would require a microscopic detail of description beyond anything implied in our ordinary specifications either of the features which make things probable or improbable or of those things which they make probable or improbable.[25] Secondly, what makes something probable or improbable need not, though it may, be anything which preceded or had any causal influence on it. The possibility that p is something that can be favoured or disfavoured, allowed or excluded, either by a prior or by a subsequent factor. This result can as easily make that condition probable as *vice versa*. Thus, a man's dissolute life may have made it more probable that he would die violently, while the trail of his blood makes it more probable that he was killed far from where he was found. Thirdly, many of our probability statements are not about the sorts of things, e.g. events, which are either determined or undetermined. Contrast, e.g., 'This is probably a wiser course than that', 'I'm probably mixing up Fermat's last theorem and Goldbach's conjecture', 'It is unlikely that the logic of chance and the logic of probability are vastly different'. It is, thus, equally mistaken either to assimilate probability to causality on the grounds that 'The anger of the workers makes a strike more likely' may be true because 'The anger of the workers has a determining influence on a strike' is true or to assimilate probability to evidence on the grounds that 'The trail of blood makes murder more probable' may be true because 'The trail of blood is good evidence for murder' is true.

Some other philosophers[26] seem to suppose that because it is the case—indeed, it is certain—that so-and-so either is or is not such-and-such, therefore it cannot be only probable that so-and-so is such-and-such. But this is a mistake, since 'It is certain that p or not-p' does not imply 'It is certain that p or it is certain that not-p'.

Probability is not a measure of the rationality of one's belief in something[27] or of the warrantability of one's assertion about it;[28] it is a measure of the relative strength of the factors in favour of or against it, the knowledge or supposition of whose existence makes it rational to believe in it. The former views of probability, like the analogous view which treats of *certainty* in terms of the reasonableness of thinking (or being certain) that something is certain,[29] put the cart before the horse. Furthermore, it may sometimes be reasonable to believe something which, unknown to one, is most improbable or to doubt something which is highly probable or even certain.

If someone has good reason for thinking X to be probable, then he has good reason to believe in and to bank or bet on X. But this does not mean[30] that for him to say that X is probable amounts to saying that he has good reason to bank or bet on it, that it is rational to believe it or that he partially believes it.[31] This is to confuse the fact that one can discover how probable someone considers X to be from the confidence of his belief in it with the supposition that this is a method of discovering how probable X actually is; a difference which is sometimes concealed by the use of the solecism 'probable for him' as a synonym for 'seems probable to him'.[32]

Subjective probability is a myth due partly to the confusion we have seen between probability and our estimate of it and partly to confusions we shall see later either between what is certain and what is known[33] or between the certainty of something and the certainty of somebody.[34]

To hold, on the other hand, that the probability of something is objective is not, as some probability theorists assume,[35] to equate its probability with the factors which make it probable, e.g. the frequency of similar occurrences in the past or the propensities of objects in the world. Such an equation would make a statement like 'Since one lecturer in ten has become a professor, the probability of a lecturer's becoming a professor is one in ten' a tautology. But it is only if and because the proportion of lecturers to professors gives the proportion of the favourable to the total factors relevant to a lecturer's becoming a professor that one can say that it gives the probability of a lecturer's becoming a professor. A statement of probability does not mention the factors which make something or its opposite probable; it says something about the degree to which these favour it and exclude the possibility of its opposite. Hence, it is a mistake to suppose that there are different senses of 'probability' or 'chance' because the grounds on which we estimate the probability or the chances of anything, e.g. the axioms of mathematics or the frequency of the occurrence of a member in a series, are different.

It is, however, just as mistaken to suppose[36] that because 'probability' does not mention the factors which make something probable, therefore 'probability' does not denote anything, as it would be to suppose that because 'influence' or 'goodness' or 'beauty' do not mention the features in virtue of which something is influential or good or beautiful, therefore 'influence' or 'goodness' or 'beauty' do not denote anything. To say that the probability of something is objective is to equate its probability with the relation between the strength of the factors favouring its being so and the strength of the factors favouring its not being so; it is to indicate how far circumstances exclude the possibility of its not being so.

A recent example of the confusion of the grounds of probability with probability itself appears among the proponents of the so called 'Propensity Theory of Probability'.[37] They do not always make it clear whether they think that the probability, or at least the chance, of something's being so is due to propensities in it and in other things or whether they think that the probability is some such propensity. To put the point in more colloquial terms: they differ as to whether what makes something *likely to* V is its being *liable to* V or whether something's being likely to V is the same as its being liable to V. And by confusing probability with propensity and likelihood with liability, the theory also confuses the possibility of 'may' with the possibility of 'can'.

NOTES

1. See the O.E.D.; cp. Greenbaum (1969), § 5.3.1.6; Moore (1962), pp. 184–8; Hacking (1967), pp. 143–68; Ayers (1968).
2. Contrast Kneale (1949), pp. 10–11.
3. Cp. 'obviously', 'conceivably', 'fortunately', 'arguably' and 'It is obvious, conceivable, fortunate, arguable'.
4. E.g. Carnap (1962); Day; Keynes, esp. p. 5; Kneale (1949); Russell; Toulmin, ch. 2; von Wright (1951c); Lucas; Mackie. Thomas Aquinas uses 'probable', in the Latin sense of 'approvable', of opinion, cp. Byrne. This is followed by Kneale and Day; contrast Black.
5. E.g. Day, p. 28; Mackie, p. 189.
6. E.g. Kneale (1949), p. 20 and Day.
7. E.g. Austin, p. 100; cp. Day, pp. 256–66 and Lucas, who also assimilate truth to certainty.
8. E.g. Toulmin and Findlay *versus* Kneale (1949) and King-Farlow.
9. E.g. Lucas.
10. Cp. Laplace. Toulmin, pp. 55, 92—cp. Findlay, p. 229 and Black, p. 465—denies this contrast because he wrongly thinks it entails, as also does Kneale (1949), who yet allows the contrast, that the probability of something is relative to the *available* evidence for it.
11. E.g. Toulmin, pp. 54–7.
12. Contrast C. I. Lewis, pp. 291, 303.
13. Contrast Keynes; Kneale (1949); Ayers (1968), ch. 2 and p. 39 and Urmson.
14. Cp. Toulmin, p. 72; contrast Kneale (1949), pp. 9–10; King-Farlow, pp. 21–2 and Keynes, p. 7.
15. E.g. Lucas, chs. 6, 11, 12.
16. *Pace* Lucas, p. 187.
17. Contrast Nagel, p. 64.
18. E.g. Hume, *Treatise* II iii 9; Carnap (1962) and Russell. In the history of language and in medieval philosophy—cp. Byrne—'probable' did have a different sense deriving from Latin 'probare'. For more

linguistic details, cp. Blom. By lumping together theories, estimates, grounds, etc., Mackie, ch. 5, manages to get at least five 'concepts' of probability.

19. E.g. Keynes and Day, p. 29.
20. E.g. Carnap (1962) and Ayer (1972), pp. 54–88.
21. E.g. Lucas, ch. I.
22. Contrast Lucas, ch. I.
23. E.g. Toulmin and Findlay; contrast King-Farlow.
24. E.g., Butler, *The Analogy of Religion* (1736), *Intro.* § 7; de Finetti, p. 147.
25. Cp. Lucas, chs. 6, 11 and 12.
26. E.g. de Finetti.
27. E.g. Bernoulli; de Morgan; Ramsey, ch. VII; Keynes; Russell; Day, p. 29; Kyburg (1961), p. 5; Ayer (1963), pp. 188–208; de Finetti. Contrast Venn.
28. E.g. Toulmin.
29. E.g. Frankfurt, pp. 317–19.
30. E.g. Ramsey; de Finetti.
31. Cp. Kyburg, ch. 3, for references.
32. E.g. Heidelberger, p. 244; cp. de Finetti, p. 102; Kyburg and Smokler, p. 6; cp. the analogous solecism 'true for him'.
33. E.g. Moore (1962), pp. 184–8 and (1959), ch. 10.
34. E.g. Frankfurt and Russell.
35. E.g. Frequency Theorists like von Mises, Carnap and Reichenbach or Propensity Theorists like Peirce and Popper.
36. E.g. Toulmin, pp. 65, ff., and Findlay, pp. 219–23; cp. Day.
37. E.g. Peirce, Volume II, § 664; Popper (1957), pp. 65–70; (1959), pp. 25–42; Hacking (1965); Mellor (1969), pp. 11–36; (1971); Levi (1967), ch. 14. Contrast White (1972).

Chapter Five

Certainty

The third member of the hierarchy, possibility, probability and certainty, has always held a dominant position in epistemology because of the opposing traditional views of it as being, on the one hand, the philosopher's chief objective and, on the other, something impossible of attainment. I shall try to show that its qualities are quite different from those often attributed to it and that they make it quite unsuitable for the role often assigned to it.

(A) THE CERTAINTY OF THINGS AND THE CERTAINTY OF PERSONS

The *certainty* of things is expressed in English in nominal, adverbial and adjectival form. Thus, we have the 'certainty' of fine weather, of the weather's being fine or that the weather will be fine. It also often happens that the weather will 'certainly' be fine. Though we do not have 'certain' fine weather,[1]* we do have 'certain' death or destruction; and it is often the case that fine weather 'is certain' or that the weather 'is certain to' be fine. But corresponding to all these ways of signifying the certainty of fine weather is the impersonal form 'It is certain that the weather will be fine'. Whenever there is a 'certainty' of something's Ving or something will 'certainly' V[2] or something is 'certain' or 'certain to V',[3] then 'it is certain that' it will V. In addition, it can be that 'It is certain that so-and-so did V or is Ving'. And *what* is certain (like what is possible or probable) when it is certain (or possible or probable) that p is not the proposition that p any more than what is suspected, feared, unfortunate or strange when it is suspected, feared, unfortunate or strange that p is the proposition that p.[4]

* References to this chapter begin on p. 88.

The certainty of persons is in many important ways different from the certainty of things. Though the certainty of something's Ving or that it will V is paralleled by the certainty of somebody's Ving or that he will V, though people, as much as things, certainly V and are as certain as things to V, and though it is as certain that people V as that things V, we do not have certain people as we do, e.g., certain death or destruction. Furthermore, people who are certain are certain in different ways from things which are certain. To say that fine weather is certain is to say that it is certain that there will be fine weather, but to say that a person is certain is not to say that it is certain that there will be a person. Things which are not certain are often probable or possible, but people who are not certain are not probable or possible either. People, but not things, can be 'certain that' so-and-so or 'certain of or about' such-and-such. People, unlike things, can also feel certain.

The certainty of persons and the certainty of things are logically independent of each other. Somebody can be (or feel) certain of something which is not itself certain, while something can be certain without anybody's being (or feeling) certain of it. 'He is certain that p' neither implies nor is implied by the impersonal 'It is certain that p'. The same thing cannot be both certain and not certain, though one person can be certain of it and another of its opposite; and it can seem certain to the one person and not to the other. People can become more or less certain of something which itself has not become any more or less certain. To express one's certainty by saying 'It is certain' is not to state that one is certain. One cannot account for the certainty of something in terms of someone's being entitled to be certain of it.[5] Someone who is (or feels) certain that p need not, logically, feel—though usually he will feel—that it is certain that p; conversely, someone who feels that it is certain that p need not feel, or be, certain that p—even though he usually will.[6] A gambler need not feel that it is certain that red will turn up next in order to feel certain that it will, nor need a sceptic, who feels certain that nothing is certain, feel, self-contradictorily, that it is certain that nothing is certain. Conversely, someone who feels, perhaps for logical reasons, that such-and-such an answer cannot be right and is, therefore, certainly not right, may not feel certain that it is not right.

I don't know exactly how a person's *being* certain and his *feeling* certain are related except that it is in the same way in which, e.g., being confident and feeling confident, being hopeful and feeling hopeful, or being afraid and feeling afraid are related and different from the way in which, e.g., being well and feeling well, being safe and feeling safe or being free and feeling free are related. There is nothing at all strange in someone's feeling well, safe or free and not being so; or being well, safe or free and not feeling so. It seems, however, difficult to be certain, confident, hopeful or afraid without feeling so or to feel so without being so; though not impossible to be in one of these states without feeling that one is in it or to feel that one is in it without being in it.

A possible difference between a person's being certain and his feeling certain is that there are criteria of rationality for the former, but not for the latter. We say 'You can't be certain', but not 'You can't feel certain', because the former properly implies a logical irrationality while the latter would improperly suggest a psychological incapacity.[7] A man who wanted to be certain before he acted would look for good grounds for believing so-and-so and not merely for something to give him a comfortable feeling.

People and things differ not only in relation to *being certain* but also to *making certain*, a difference which is often reflected in the distinction between *making certain* and *making something certain*. *To make something certain* is to make certain or to make it certain that there *be* so-and-so. *To make certain*, on the other hand, is to make certain that there *is* or was or will be so-and-so. Making certain is akin to ascertaining: making something certain to ensuring. Both people and things can make something certain, as when the collapse of the economy makes ruin certain, the newly discovered letters make it certain that Tennyson was the author of that poem, or when, by precautions, I make certain that no one will follow me. But only people can make certain, as when by investigations I make certain that no one has been following me. The weather can make it certain, but not make certain, that there will be no further play. To make something certain is to *remove* the possibility of any alternative to it, whereas to make certain is to *reveal* the absence of any such possibility.

Someone or something can make something probable or unlikely, but no one—nor, of course, nothing—can make probable or unlikely. We can make or make something absolutely or completely certain. But we can only make something more or less certain; we cannot make more or less certain. But, whether someone makes certain that p or someone or something makes it certain that p, it follows that it is certain that p.

Different again is the notion of something's making *somebody* certain. Making *somebody* certain, unlike making *something* certain, is not making it certain that there be somebody, nor is it making certain that there is somebody. It is making somebody certain that so-and-so. But to make somebody certain that so-and-so is not to make, or to make it, certain that so-and-so, any more than assuring somebody that p is either ascertaining or ensuring that p. It would be a poor joke to say that someone had made certain—much less had made it certain—that no one would follow him because something had made him certain of this. One cannot have made certain or made it certain that p unless it is certain that p. Conversely, if somebody makes certain that so-and-so or makes it certain that so-and-so, this is not necessarily to make himself certain. For he could be certain because he wrongly thought that he had made certain or made it certain, and he could fail to be certain because he did not realize that he had made certain or made it certain.

(B) CERTAINTY AND POSSIBILITY

'Certain',[8] as its etymology suggests, means settled. If something is certain, it is settled because there is no possibility of its not being so. If somebody is certain of something, he is settled because he does not entertain the possibility of its not being so; perhaps because he thinks there is no such possibility. To make something certain is to remove any possibility of its not being so; to make certain is to reveal the absence of any such possibility. People can do both, but things can do only the former.

The possibility which is excluded by certainty is the possibility of 'may', not the possibility of 'can'; that is, it is the possibility that something *is* (or was or will be) so, not the possibility that it (*should*) be so or the possibility *for it to be* so. We have seen that it is quite appropriate to say 'He could have Ved (or "It was possible for him to have Ved"), but he certainly did not'; whereas it is absurd to say 'He may have Ved (or "It is possible that he did V"), but he certainly did not'. Likewise, it is because 'He possibly Ved' means 'It is possible that he Ved' and not 'It was possible for him to V' that it is equally absurd to say 'He possibly Ved, but he certainly did not'.

It is because the possibility which is excluded by certainty is not the possibility of 'can' that the typical evidence of the sceptic to show that something *could* either logically or empirically be other than it is thought to be does not put in doubt the certainty that it is what it is thought to be. What the sceptic would need to provide is evidence that something *may* (or might) be other than it is thought to be. And it is because the possibility excluded by certainty is the possibility which goes with probability that 'possible, probable and certain' form a hierarchy and that light thrown on the notion of certainty illuminates the notion of probability.

A change in evidence can make something seem certain (or probable) which before seemed uncertain (or improbable), and *vice versa*; but only a change in circumstances can alter the actual certainty—or probability—of anything. Hence, while the certainty (or probability) of things both in the future and in the past can seem different from what they are, and while the circumstances which make something certain can be prior or subsequent to it, only things in the future can become certain or uncertain (probable or improbable). This is why we can say, e.g., 'In January it *was* not certain (very probable) that Britain *would* apply for entry into the Common Market, but by June this *had become* certain (very probable)', but not 'In June it *was* not certain (very probable) that Britain *had* applied for entry into the Common Market, but by June this *had become* certain (very probable)'. We can also, of course, say 'In June it *seemed* certain (very probable) that Britain *had* applied for entry into the Common Market, but this was a mistake'. But we have to distinguish 'It was a mistake to say (or think) in June that this was certain (probable, sad, surprising)' and 'It was not certain (probable, sad surprising) in June', for the contrast is between something in the past's *being* certain (probable, sad, surprising) and its *being said* or *having been said* to be so, and not between its *having been* and its *having been said* to be so.

(C) CERTAINTY AND KNOWLEDGE

Some philosophers have identified knowledge with certainty, either in the form that *knowing something* is alleged to be the same as *being certain* about it or in the quite different form that *what is known* is alleged to be the same as *what is certain.* I wish to dispute both of these equations.

Any identification of *knowing* that p with *being certain* that p must be mistaken, for the one neither implies nor is implied by the other. A man can be (or feel) certain that p without knowing that p, still more without knowing that he knows that p,[9] and, indeed, even without being right in supposing that p. One man can be certain that p and another be certain of the opposite, but if somebody knows that p, no one can know the opposite. Contrariwise, a man can know that p without being (or feeling) certain that p[10] and without thinking or feeling that p is certain, though not without being right in supposing that p. Indeed he may well know something without either knowing or even thinking that he does know it.

The notions of *knowing* that p and *being certain* that p are, indeed, quite different. Knowing, unlike being certain, implies being right. One can sound or look certain, but there is no physical expression of knowledge. Certainty, not knowledge, is something one can feel, something one can induce in oneself or another, something one can discover in oneself introspectively or in another behaviouristically. We can ask *how* somebody knows that p, but not how he is certain that p; and, conversely, *why* he is certain, but not why he knows. To ask how certain he is is quite different from asking how he is certain and also from how knowledgeable he is. One can argue that if somebody knows that p, then it is known that p and *vice versa*, but not that if somebody is certain that p, then it is certain that p, or *vice versa.* This is why we have the question 'To (or by) whom is it known?', but not the question 'To (or by) whom is it certain' and the answer

'It is well (widely, universally) known', but not the answer 'It is well (widely, universally) certain'.

One source of the confusion[11] about *knowing* and *being certain* may lie in the fact that one commonly *says* 'I know' when one is certain, and *vice versa.* But the appropriateness of saying 'I know' must not be confused with its truth. And, as we have just seen, there is no implication by *knowing* of *being certain.* A second source of the confusion may spring from the connection of knowing or not knowing what to do with being certain or uncertain about one's plans or intentions. Not knowing what to call one's dog is the result of indecision or uncertainty; all that one needs for knowledge here, unlike knowing what one's neighbour's dog is called, is to make up one's mind. But it is the end of the uncertainty that gives us the knowledge, and not *vice versa.* Contrariwise, a third source of the assimilation of the two notions may be that coming to know something can, though it need not, end one's uncertainty about it, whether what one comes to know is what to do or what is the case.

A fourth source of the confusion between somebody's knowing and being certain and possibly also of that between something's being known and being certain is the assimilation either of *knowing* or of *being certain* to *knowing for certain.*[12] But, in fact, 'know for certain' is neither a redundant form of 'know' nor an equivalent of 'be certain' nor a conjunction of the two.

'Knowing for certain' cannot be the same as 'knowing and being certain', since 'for certain' here means, as it does also in 'tell, establish, show, discover for certain' that something is certain, whereas 'being certain' here means only that somebody is certain. One could know and be certain that p even though one did not know for certain that p. Yet, although 'for certain' implies that something rather than someone is certain, 'know (can tell, establish) for certain' does not mean[13] that it is certain that someone knows (can tell, establish) that is, that someone certainly knows (can tell, establish) but that someone knows (can tell, establish) in such a way that what is known (told, established) is revealed as certain.

Nor can *knowing for certain* be the same as *being certain. Knowing for certain* implies *knowing* and, therefore, being right, whereas we have seen already that one can be certain without knowing or even being right, and also know without being certain.

Nor, finally, can *knowing for certain* be the same as simply *knowing*, because, although the former implies the latter, it is not implied by it. To know for certain, like to tell, establish, discover, make out for certain, is to know in such a way that the possibility that one is mistaken is excluded. If, on the other hand, one simply knows, then though it follows that one *is not* mistaken, it does not follow that *it is not possible* that one is mistaken.[14] A failure to distinguish these two is easy if in the former one replaces 'follows' by 'cannot be that not', thus giving the correct assertion 'If one knows, it cannot be that one is not right'. But a similar replacement in the latter form then gives us the quite different and incorrect assertion 'If one knows, it cannot be that it is possible that one is not right'. Even if, as I shall dispute below, 'It is known that p' implies that it is certain that p, it cannot be concluded that 'It is known that p' implies that it is known for certain that p; for 'It is certain that p', as we shall see, does not imply that it is known that p, much less that it is known for certain that p. If it is certain that p, then there must exist circumstances which either make it certain that p or would show that it is certain that p, but someone who knows that p need not know of the existence nor appreciate the force of these circumstances: and, therefore, need not know for certain that p. Sometimes, having discovered that p—e.g. that there are no survivors, that the bus goes at 8 or that the new drug cures the disease—we feel impelled to take further measures to make certain that p, that is, measures which enable us to exclude any other possibility—e.g. that a wounded passenger managed to crawl into the jungle, that the time-table is different on Good Friday or that the drug merely suppresses the symptoms.

Having seen that somebody's knowing that p and somebody's being certain that p are so far from being the same that neither implies the other, let us look at the alleged identification[15] of *it's being known* that p with *it's being certain* that p, that is, of what is known with what is certain.

This equation too must be wrong, for what is certain need not be known. It could well be certain that the earth will be desolated by solar radiation before A.D. 2000, although no one knows or even thinks this. The circumstances which make something certain need not make it known. Nor need the evidence which reveals the non-existence of the possibility that something is not so be appreciated as such by those who have this evidence. Furthermore, if whatever was certainly so was known to be so, then whatever was necessarily so would also have to be known to be so, since whatever is necessarily so is certainly so. But there can be many necessary results which are not known.

The view, expressed by G. E. Moore,[16] that 'a thing can't be certain unless it is known' depends—as he later partly saw[17]—on assimilating this mistaken view to the plausible view that no one would, or could appropriately, *say* that something was certain unless he thought, and was therefore prepared to assert, that he, or somebody, knew it.

Similarly, though 'The possibility that not-p is not excluded' follows from 'It is not certain that p', it does not, as Moore thought,[18] follow from, much less is it logically or in meaning equivalent to, 'Nobody knows (or knows for certain) that p'. It is because Moore thought that what is certain is relative to somebody's knowledge that he also, wrongly, held[19] that 'Two different people who say at the same time about the same proposition p, the one "It is certain [or 'possible', as he says in another passage] that p is true", the other "It is not certain [or 'possible'] that p is true", may both be saying what is true and not contradicting one another.'

Since the mistaken thesis that if something is certain it must be known is equivalent to the thesis that if it is not known, it is not certain and, therefore, the possibility of its opposite is not excluded, to say that if it is not known that p, then it is possible that not-p is also mistaken. Equally mistaken is a recent variation[20] on this view, according to which 'a state of affairs is possible if it is not known not to obtain, and no practicable investigations would establish that it does not obtain'. If this were correct, it would follow, by contraposition, that if this state of affairs is not possible, then either it is known, or practicable investigations would establish, that it does not obtain. But there is no reason at all to suppose that nothing can be certain unless practicable investigations would establish the non-existence of its opposite. Such a view borrows part of its plausibility from the fact that a reason for supposing something to be certain is a supposition that its opposite can be practically disproved and partly from the fact that if something is certain, then its opposite does not exist.

Having established that what is certain need not be known, let us inquire whether what is known need be certain.

First, this view may rely on the argument that because it is certain that, if it is known that p, then p—which itself follows from the fact that it is necessarily true that, if it is known that p, then p—therefore, if it is known that p, it is certain that p. But the principle underlying this argument, namely, that if it is certain that if p, then q, then if p, it is certain that q, is as mistaken as the analogous principle that if it is necessarily true that if p, then q, then if p, it is necessarily true that q. For, according to it, from the correct premiss that it is certain (because necessary) that if p, then p, there would follow the mistaken conclusion that if p, then it is certain that p. What both arguments do is to argue from the necessity or certainty of the connection between p and q—the medieval 'necessitas consequentiae' —to the necessity or certainty of q itself—the medieval 'necessitas consequentis'; that is from 'It is certain (necessary) that it follows' or 'It certainly (necessarily) follows' to 'It follows that it is certain (necessary)'. 'If p, then certainly q' no more means 'If p, it is certain that q' than 'If p, then obviously, clearly, evidently, probably q', means 'If p, it is obvious, clear, evident, probable that q'. For if they did, the contrapositives would be 'If it's not certain, obvious, clear, evident, probable that q, then not-p'; but the actual contrapositives are 'If not q, then certainly, obviously, clearly, probably, evidently not p'. Hence, the fact that it is certain that if it is known that p, then p, no more provides a reason for supposing that if it is known that p, then it is certain that p, than the fact that it is necessary that if it is known that p, then p provides a reason for supposing that if it is known that p, then it is necessary that p.

Secondly, the view that what is known must be certain may depend on wrongly assuming that, because some ways in which I come to know something are such that I thereby make certain of it, therefore, whenever and however I come to know something, I thereby make certain of it and so reveal that it is certain. But this is to confuse *know* and *know for certain*.[21] It is true that if something is known for certain, then it is certain, for to know for certain is to know in such a way that the possibility that what is known to be so might be other than it is, is excluded. But it does not follow that because its being known for certain that p implies that it is certain that p, therefore its being known that p implies that it is certain that p.

Thirdly, this view, beloved of sceptical philosophers,[22] limits the area of knowledge in a quite implausible way. People often claim, and rightly, to know things which are not certain. Furthermore, if what is correctly believed can, without further change in it, become known, then, on the assumption under discussion, only what is certain could be correctly believed. And this is false.

Finally, if whatever is known is therefore certain, the interesting, and perhaps awkward, consequence follows that if there is anything which cannot be so without being known to be so, then it is also something which cannot be so without being certain. In other words, it is something for which the generally invalid inference 'p implies it is certain that p'[23] would be valid. For instance, if a man in full possession of his senses cannot help knowing that he is alive and knowing whether and what he is feeling and thinking, then if he is alive and feeling and thinking so-and-so, it is certain that he is. And if God knows all, then all is certain.

G. E. Moore[24] on one occasion tried to avoid this conclusion by arguing that for the truth of 'It is certain that p' 'I know that p' is sufficient, but not necessary, while 'Somebody knows that p' is necessary, but not sufficient. Thus, on the one hand Hitler's knowledge that he is alive would not show that he is certainly alive, but mine would; while on the other, for it to be certain that he is dead requires somebody's knowledge of it, though not mine. This attempt to escape from his own dilemma will not, however, work. The second alternative implies the conclusion, which we have already seen to be false, that what is certain must be known; while the first alternative commits us to the debatable conclusion that by knowing that p I make certain that p.

Our conclusion, therefore, is that neither need what is certain be known nor need what is known be certain; and, hence, that *what is certain* and *what is known* are not the same. Furthermore, even if whatever is known were certain, this is not the only way in which something can be certain. Another way, as we have seen, is by its being necessary; for whatever is necessarily so is certainly so. If the light cannot have switched itself on, then it is certain that it did not. The conceptual connection between certainty (or the possibility of 'may') and knowledge is no closer than that between certainty (or the possibility of 'may') and necessity.

The philosophically popular name of 'epistemic possibility'[25] is a misnomer and the ideas based on it are mistaken.

(D) CERTAIN AND SURE

Philosophers usually speak as if 'certain' and 'sure' were synonymous, at least in such phrases as 'A is (or feels) certain that p' and 'A is (or feels) sure that p'.[26] The two notions do, indeed, show many similarities and may sometimes even be used rather indifferently in common speech. One can be, feel or make either certain or sure. And one can make sure, as one can make certain, either by prevention or disproof of the possibility of the opposite. He—and, doubtfully, it—can be either certain or sure that p and either certain or sure to do X. One can be fairly, pretty, quite, absolutely or perfectly certain or sure. For both concepts, degrees of approximation, e.g. quite or almost, may sometimes be expressed as if they were degrees of intensity, e.g. very or more. Feeling sure, like feeling certain, is a completeness feeling consisting in the absence of its opposite; and feeling sure has to being sure the same sort of relation as feeling certain to being certain.

Furthermore, the two notions seem, at least often, to be mutually implicative. If he is certain or it is certain that p or certain to do X, then he is sure and, perhaps, it is sure that p or to do X; and *vice versa.* Many things which are certain, e.g. death, ruin or success, are also sure, and what is a sure sign of or means to something is also a certain sign of or means to it. The two notions have similar relations to knowledge: a person's being or feeling sure that p no more implies or is implied by his knowing that p than does his being or feeling certain that p.

Despite these similarities, however, the two notions are different; though related in a way which explains the similarities. Just as 'certain' retains its etymological suggestion of 'settled'

(Latin 'certus'), so 'sure' retains its etymological suggestion of 'free from care or worry' (Latin 'securus'). Its basic use is in the personal 'He is sure that p'. If somebody or something is sure to do X, then one can be sure that he or it will do X. This is why 'It is sure that p' is rare or non-existent and why 'making it sure that p', unlike either 'making sure that p' or 'making certain that p' or 'making it certain that p', seems not to exist. To call an object 'sure' is to imply that it has characteristics which allow a person to be sure about it; a sure sign, criterion, or means, a sure income, step or foothold is one of which someone can properly be, even if he is not actually, sure. This is why a step, foothold, niche or refuge is commonly called 'sure', but not 'certain'.[27] One can feel sure of oneself, but not certain of oneself. 'Certain' is basically related to the exclusion of some possibility; 'sure' to the exclusion of some dubitability. 'Senate will *surely* not agree to that' suggests, in a rhetorically interrogative way, that there is no doubt; 'Senate will *certainly* not agree to that' states categorically that there is no possibility. 'Surely, they will agree' means 'Is it not safe to say that they will agree?', whereas 'Certainly, they will agree' means 'It is definite that they will agree'. This is why 'Surely, he may use your name' is appropriate where permission is being asked, but only 'Certainly, he may use my name' where permission is being granted.

The close connection of 'certain' with 'sure'—even their mutual implication—is explained by the particular kind of worry whose absence 'sure' signifies. It is the worry that something may or might be other than it appears to be or become other than it is. To be (or feel) sure that p is to be (or feel) free from any worry that the possibilty of not-p has not been excluded; to be (or feel) certain that p is to be settled in the opinion that the possibility that not-p has been excluded. Hence, one cannot be (feel) sure unless one is (feels) certain; *and vice versa.*

Because philosophers have usually assumed that to be (or feel) sure is the same as to be (or feel) certain, the differences between them have not been explored. The only exception I know is the suggestion by J. L. Austin,[28] perhaps made only in passing, that ' "Being certain" tends to indicate confidence in our memories and our past discernment, "being sure" to indicate confidence in the current perception.' But this suggestion does not square with the normal use of the terms. My past experience of colours could well make me (feel) sure that these curtains will fade in strong sunlight and my present perception of this material make me certain that its colour does not match that of the cushion covers. The fact is that it is not what makes a man certain that is different from what makes him sure; what is different is what each of these makes him.

The philosophical implications of the rather subtle, but I believe definite, difference between 'certain' and 'sure' should not be over-emphasized; though its neglect may be partly responsible for the common misleading linking of certainty and indubitability in preference to certainty and impossibility, and, hence, for the supposition that reasonable doubt about something makes that thing uncertain.

NOTES

1. The use of 'certain'—and the equivalent words in other languages—as 'particular' is quite different. A certain building is not a building which is certain.
2. The correspondence of the forms 'certainly' and 'It is certain that' has many parallels, e.g. possibly, probably, conceivably, arguably, definitely, clearly, unfortunately, surprisingly.
3. The correspondence of the forms 'certain to' and 'It is certain that' has many parallels, e.g. likely to/that, expected to/that, agreed to/that.
4. Contrast Rollins, pp. 67–71.
5. Contrast Frankfurt, p. 319; Hamlyn, pp. 517–18.
6. Compare 'hopeful', 'doubtful', 'dubious', 'sure'. In general, to feel F that p does not imply that there is a form 'to feel that it is F that p', since the former may make sense while the latter does not, e.g. 'afraid', 'confident', 'glad', 'sorry'.
7. Compare 'You *can't* be sure, confident', 'You *can't* expect, believe'.
8. 'Certain' is qualified by 'fairly', 'quite', 'more', 'reasonably' 'absolutely', 'completely', 'pretty', 'nearly', 'practically', 'not at all' in the same way as are, e.g., 'clean', 'complete', 'correct', 'free', 'full', 'new'.
9. Contrast Chisholm (1957), p. 19.
10. Contrast Locke, *Essay*, IV xi 9; Moore (1953), p. 273; (1959), p. 239; Malcolm (1963), pp. 59, 244; Ayer (1956), ch. I; Aaron, pp. 1–13.
11. E.g. Wittgenstein (1969), § 8; contrast, § 308.
12. E.g. Moore (1959), pp. 227, 236, 239; (1953), p. 273; Malcolm (1963),

pp. 25 and 32; contrast p. 229; Hintikka (1962), pp. 116–22; Chisholm (1957), p. 19.

13. Contrast 'He'll come for certain', which does mean 'He'll certainly come', that is, 'It is certain that he will come'.
14. As Wittgenstein (1969), § 15, seems to have thought.
15. E.g. Moore (1962), pp. 184–8. This is often called 'epistemic certainty'.
16. (1959), p. 240; (1962), pp. 277–9; (1912), p. 136; cp. Rollins.
17. In (1959) he sometimes said that 'It is certain that p' implies '*I* know that p', but in (1962), p. 277 he moved to the view that 'It is certain that p' implies '*Somebody* knows that p'. I argue that both views are wrong.
18. E.g. (1962), p. 184; (1959), pp. 197, 205 *et passim.*
19. E.g. (1959), pp. 235, 241; (1962), pp. 277–9; cp. the philosophical barbarism 'certain for (or to) somebody', e.g. Heidelberger (1963), p. 243; Firth (1967), pp. 3–27.
20. Hacking (1967).
21. As Moore sometimes did, e.g. (1959), pp. 220, 222, 238–41.
22. E.g. Heidelberger (1963), pp. 243–4; cp. Robinson (1971), p. 22.
23. Incidentally A. Naess's opinion poll method of doing philosophy showed that the vast majority of his subjects thought that 'p' is indeed equivalent to 'It is certain that p' as well as to 'It is true that p'.
24. E.g. (1962), p. 278; cp. (1959), pp. 240–1.
25. E.g. Moore; Hacking; Heidelberger, *op. cit.*; cp. Russell (1956), pp. 254–5; Ayers (1968), ch. 2; Gibbs, pp. 344–5; von Wright (1951b), p. 32.
26. E.g. Moore (1959), p. 238; Chisholm (1957), p. 19.
27. We have 'quite certainly', but not 'quite surely'; 'surely enough', but not 'certainly enough'.
28. (1961), pp. 58–62.

Chapter Six

Necessity

(A) NECESSITY

The general notion of *necessity* is so pervasive in our thinking that it takes widely different forms, such as those expressed by 'necessity', 'must', 'has to', 'needs', 'bound to' and 'obliged'. Each of these different forms repays, and will be given, separate treatment. For instance, 'must' conceals an interesting and important difference between the subjunctive-governing and the indicative-governing use, which we have already seen in the case of 'possible' and will see again in that of 'ought'. The confusion of 'need' with 'want' and of 'obliged' with 'ought' is the source of many important philosophical puzzles.

Let us, however, start with the general notion of *necessity*, which, like that of *possibility*, can be expressed either nominally ('necessity'), adverbially ('necessarily') or adjectivally ('necessary'), though we shall see that, also like *possibility*, all three grammatical forms can be explained in terms of the adjective 'necessary'.

Necessity is something one can see, feel or discover, anticipate, meet or avoid, consider, deny or rule out; it is something one can be confronted with, delayed, or thwarted by. It can be financial, economic, practical or logical. It can be the necessity *of* something (e.g. agreement, delay or expense), *of* doing something (e.g. travelling, sleeping or eating) or *of* something's happening (e.g. the pipes unfreezing, the kettle's boiling); it can be the necessity *to* do something (e.g. to wait, to move quickly) or the necessity *for* something (e.g. further delay, fresh supplies). Unlike problematic possibility (i.e. the possibility expressed by 'may'), probability or certainty, there is no necessity *that* something *is* so, but only the necessity—like the existential possibility (i.e. the possibility expressed by 'can')—*that* something (*should*) *be* so or *for* it *to be* so. Something can happen or be done *of* or *from* necessity, but not of or from probability and certainty. The necessity of something's occurrence, unlike the probability or certainty of its occurrence, can explain its occurrence; something can happen because it is necessary, but not because it is probable or certain. But whenever there is, in any of these ways, a 'necessity'—financial, practical or otherwise—for something, then 'It is necessary' for that to be.

The necessity of what is expressed by a verb or adjective is indicated by an accompanying adverb 'necessarily'; e.g. 'Unemployment necessarily results in suffering', 'All of us necessarily share the blame', 'His punishment was necessarily harsh and long'. 'Necessarily' can also be used as an intensifier with verbs, such as 'have to' and 'must', which themselves signify necessity. Contrariwise, the negative 'not necessarily' signifies that there is no necessity for that which is expressed by the accompanied verb or adjective; e.g. 'This will not necessarily solve any problems', 'It does not necessarily mean what it says', 'A replacement is not necessarily expensive'. The qualifier, 'not necessarily', which means 'It is not necessary for so-and-so to be' is quite different from the qualifier 'necessarily not', which means 'It is necessary for so-and-so not to be'. This is why 'possibly, but not necessarily' makes sense, whereas 'possibly, but necessarily not' is a contradiction. The adverb 'unnecessarily' is a further negative quite different both from 'not necessarily' and from 'necessarily not'. To say of anything that it is 'unnecessarily' offensive, obscure or expensive is to say quite a different thing from saying that it is 'not necessarily' or 'necessarily not' offensive, obscure or expensive. It need not be that what is not necessarily so is so and it cannot be that what is necessarily not so is so; whereas it must be that what is unnecessarily so is so, even though it is not something necessary. We can use 'not necessarily' or 'necessarily not' of anything, but there are many things to which it would be unintelligible to attach 'unnecessarily'. X may properly be said to be not necessarily or necessarily not a good thing, the Prime Minister, true or valid; but it couldn't sensibly be said to be unnecessarily a good thing, the Prime Minister, true or valid.

Whatever can be expressed with the adverb 'necessarily' or, as we saw, with the noun 'necessity', can be expressed with the adjective 'necessary'. If unemployment necessarily results in suffering, then it is necessary for unemployment to result in suffering; if someone's punishment was necessarily harsh and long, then it was necessary for it to be harsh and long. Likewise, if something is not necessarily so, it is not necessary for it to be so; if it is necessarily not so, it is necessary for it not to be so and if it is unnecessarily so, it is unnecessary for it to be so.

Necessity can be either prospective or retrospective, that is necessary *for*, or as a means to, something else, or necessary because *of*, or as a consequence of, something else.[1]* Thus, a particular condition may be necessary for a particular result and a particular result necessary because of a particular condition. To ask why something is necessary may be to ask what it is necessary for or what makes it necessary. In relation to its other, it is necessary for one thing to be; that is, it is not possible for it not to be. Something can be necessary for one thing, but not necessary or unnecessary, for another; it can be necessary because of one thing, but not necessary because of another.

Necessity, whether retrospective or prospective, is relative. In a given set of circumstances, something will be necessary either because something else is so or in order for something else to become so. Furthermore, it will be thus necessary only under a certain aspect, which is indicated by some such qualification as 'logically', 'legally', 'physically', 'practically', 'in theory', etc.[2] It is 'necessary' because in these circumstances and with either this antecedent or this prospect it is, as far as this aspect of the matter is concerned, the *only* possibility. It is impossible under these conditions for there to be anything else. There not only is not, but there could not be, any alternative.

(B) NECESSARY TRUTH

Before examining the various varieties of necessity, it is worth digressing shortly to consider a curious anomaly in philosophical talk about necessary truths. Truths share with results, con-

* References to this chapter begin on p. 97.

sequences, inferences, connections and conditions, but not with journeys, deaths, apologies or expenses, the characteristic that they can be not necessary without being unnecessary. Necessary truths are contrasted with not necessary truths, but not with unnecessary truths.[3] Furthermore, a necessary truth, e.g., in logic or mathematics, is not necessary *for* something;[4] it is necessary *because of* something.

There is, moreover, a curious anomaly in the philosophical phrase 'necessarily true' which can be brought out as follows. Qualifying the existence of a quality Q clearly differs from qualifying the quality Q itself; the one is often expressed by saying that something would 'Mly be Q', the other by saying that it would 'be Nly Q'. For example, one might say that something would fortunately, apparently, luckily or rarely be true, spoilt or safe; this would be quite different from saying that something would be wholly, partly or entirely true, spoilt or safe.

'Necessarily', like 'not necessarily' and unlike 'unnecessarily', seems most commonly to qualify the existence of a quality, as when it is said that handmade shoes would necessarily be expensive or a first edition necessarily be rare or a prime number necessarily be unfactorizable and a particular statement necessarily be unprovable. In the same way, it can often be said of a statement that it would necessarily be true. For example, if it were true that all men are mortal and that Socrates is a man, then it would necessarily be true that Socrates is mortal. On the other hand, philosophical talk of 'necessary truths' or of 'what is necessarily true' cannot mean that something would 'necessarily be true' but that it would 'be necessarily true', for its contrast of 'contingent truths' or of 'what is contingently true' is not a contrast of what would 'not necessarily be true' or of the ungrammatical 'what would contingently be true', but of 'what would be contingently true'. Philosophers cometimes assert that so-and-so could be or could not be necessarily true, but it would be nonsense to argue whether anything could or could not necessarily be true. Furthermore, to say that in certain conditions it would 'necessarily be true' that Socrates is mortal is not to say that it would ever 'be necessarily true' that Socrates is mortal.

The philosophical phrase 'necessarily true' is not, therefore, exactly analogous *either* to 'necessarily expensive'—since it has no contrast 'unnecessarily true' to compare with 'unnecessarily expensive'—*or* to 'necessarily lost'—since the expansion of 'necessarily true' is 'be necessarily true' whereas the expansion of 'necessarily lost' is 'necessarily be lost'.[5]

To say, therefore, that it is necessarily true that X is Y—which is logically equivalent to, though not the same as, saying that X is necessarily Y—is to use 'necessarily' in the ordinary way in which it is used in saying, e.g., that it is necessarily difficult to travel without a passport. Both 'It is necessarily true that X is Y' and 'X is necessarily Y' are quite different[6] from the usual philosophers' assertion that *X is Y* is a necessary truth. The latter implies the former, but not *vice versa.* The connection between *being necessarily true*, that is being a necessary truth, and *necessarily being true*, the former of which implies, but is not implied by, the latter, may be that for something necessarily to be true is to have its truth necessitated by something else as when it would necessarily be true that Socrates is mortal if Socrates were a man and all men were mortal, whereas for something to be necessarily true is to have its truth necessitated by itself as when it would be necessarily true that all men are male. Alternatively, what would necessarily be true would be true given some particular circumstances, whereas what would be necessarily true would be true whatever the circumstances. A confusion of these two is a source of the view, held by Plato, Hume and the Logical Positivists, that the truth of 'If p is known to be true, then p is necessarily true' shows that only what is necessarily true can be known. It is also, as we shall see in the final chapter, a cause of confusion about the nature and implications of modality.[7]

(c) NECESSITY AND CERTAINTY

Necessity and certainty are contrasted with different kinds of possibility. The former excludes a possibility expressed by 'can', whereas the latter excludes a possibility expressed by 'may'. If it is necessary to V, then it is not possible not to V, and if it is necessary that it should V, it is not possible that it should not V; whereas if it is certain that it Vs, it is not possible that it does not V. Hence, whereas something can be logically, physically financially, politically or legally necessary in contrast to being in one or other of these ways (existentially) possible, it cannot be in any of these ways certain, probable or (problematically) possible.

Necessity and certainty are, indeed, quite different. It can be certain that, but not necessary that, something *is*, *was* or *will be* so; and necessary that, but not certain that, something (*should*) *be* so. Somebody or something can be certain to V, but not necessary to V; whereas, in the impersonal use of 'it', 'It is necessary to V' makes sense, but not 'It is certain to V'. If somebody or something is certain to V, it is certain that he or it *will* V; whereas if, impersonally, it is necessary to V, then it is necessary that there *should* be Ving. Thus, to say 'He is certain to be late' or 'It is certain that he will be late' makes sense, but not 'He is necessary to be late' or 'It is necessary that he will be late'. On the other hand, to say 'It is necessary to say our prayers' or 'It is necessary that we should say our prayers' makes sense, but not 'It is certain to say our prayers' or 'It is certain that we should say our prayers'. It can be necessary, but not certain, to V in order to F. Something can be necessary for one thing or person and not necessary (or unnecessary) for another; but it makes no sense to talk of anything being certain or not certain (or uncertain) for something or somebody, nor, therefore, of its being certain for one and not for another. Necessity also, unlike certainty, can explain. Something can happen because it was necessary for it to happen, but not because it was certain that it would happen.

Clearly, since the range of the necessary covers the possibility expressed by 'can', while that of the certain is restricted to the narrower possibility expressed by 'may', any relationship between certainty and necessity can only hold within the area of the latter possibility. There can be no question of whether what is financially, politically or legally necessary is or is not in these ways certain, since there is no such thing as these different kinds of certainty. Since also there is no implication between what is financially, politically or legally necessary and what is logically or physically necessary, there can be no implication between what is financially, politically or legally necessary and what is simply certain.

A clue to the relationship between certainty and logical or physical necessity lies in the double use of *necessity*, either prospectively, that is, necessary *for* so-and-so, or retrospectively, that is, necessary *because* of so-and-so. A particular condition may be necessary for a certain result and a particular result necessary because of a particular condition, as when a rest is necessary for further effort or necessary because of past effort.

Certainty does not entail necessity, either prospective or retrospective. Many a certain death has not been necessary for any purpose or because of anything that made it certain. Most people suppose that continued inflation is certain, but few of them believe that it is necessary either for anything or because of anything. You can certainly agree with me without necessarily agreeing with me. Even in the artificial formulae of logicians, 'p is certain', which implies that not-p is not possible (= may), does not imply 'not-p is not possible (= can)', and, therefore, does not imply the latter's equivalent, 'p is necessary'. What is certainly true need not either necessarily be true or be necessarily true.

Whether, on the other hand, necessity entails certainty depends on whether the necessity is prospective or retrospective. Since what is prospective need not come to pass, what is necessary for it may not occur and cannot, therefore, be certain. Most people think that a halt to inflation is necessary—that is, necessary for the good of the country—but hardly anyone seems to think that it is certain. We can all think of many necessary improvements, very few of which are certain to be made. On the other hand, what is necessary because of, or what is necessitated by, something else has its occurrence fixed by this and is, therefore, certain. If continued inflation is necessary because of the nature of the economy, then it is certain. Since the adverbial form 'necessarily' seems always to signify retrospective necessity, what is necessarily so is certainly so. If disappointment necessarily awaits us, then it certainly awaits us. Similarly, what would necessarily be true—and what would be necessarily true—would certainly be true.[8] What makes it necessary for so-and-so to be true thereby makes it certain that it is true.

The certainty of what is certain because it is necessary is no different from the certainty of what is certain though it is contingent; the difference is simply in the reasons for the certainty. The once fashionable distinction between the certainty of the truth of non-empirical statements and that of empirical statements is simply a mistake. Equally mistaken is the sceptical thesis which confuses the correct view that no empirical statement is necessarily true—though such a statement can, in certain circumstances, necessarily be true—and the incorrect view that no such statement is certainly true. It can be quite certain, and a person can properly be quite certain, that p, even though it is possible for it to be that not-p.

In the next three chapters, I shall examine the particular nuances of 'necessity' expressed in the verbs 'must' (and 'have to'), 'need' and 'be obliged'.

NOTES

1. Harré, p. 44, supposes that all necessity is 'necessity for'.
2. None of these features of necessity implies that there is more than one meaning of 'necessary'; *pace* Braine, pp. 139–70.
3. Cp. the philosophical phrase 'necessary proposition'.
4. *Pace* Hamlyn.
5. We also say 'would (not) necessarily be so' rather than 'would (not) be necessarily so'.
6. *Pace* Quine (1960), § 41; (1953), ch. 8.
7. Contrast Quine *op. cit.* and Plantinga, pp. 235–58.
8. Austin's apparent denial of this (1962), p. 117, depends on his linking of 'certainty' with 'firmly established'.

Chapter Seven

Must

'Must' can accompany almost any verb to indicate either subjunctively that it must be that something (should) *be* so, e.g. 'He must stop doing that', or indicatively that it must be that something *is* so, e.g. 'He must be in the other room'. Because almost any example of 'must V' could with sufficient ingenuity be taken either way, only the context shows which is meant, e.g. 'He must work harder than his colleagues, if he is to finish in time' contrasted with 'He must work harder than his colleagues, if he always finishes so early' or 'You must be bold, if you are to succeed' contrasted with 'You must be bold, if you did succeed'.

The distinction between the subjunctive-governing and the indicative-governing uses accounts for various differences in the behaviour of 'must'. Thus, the subjunctive is neutral about whether what must be is already so, e.g. 'He must work harder than his colleagues (if he is to finish in time)', whereas the indicative implies that what must be is so, e.g. 'He must work harder than his colleagues (if he has finished so early)'. 'He will have to work harder than his colleagues' is an alternative form for the subjunctive use, but not for the indicative use, of 'He must work harder than his colleagues'. Similarly, the past form of the subjunctive-governing use is 'had to', e.g. 'He had to work hard when he was young', whereas the past form of the indicative-governing use is 'must have', e.g. 'He must have worked hard when he was young'. And, whereas 'must have Ved' implies 'did V', 'had to V' does not necessarily do so. If it is true that he must have worked hard, then it follows that he did, whereas if it is true that he had to work hard, it only follows that he did if that in virtue of which he had to work was a physical necessity. It could well be that legally or contractually he had to deliver the goods within a fortnight, but that he failed to fulfil his contractual obligations. Furthermore, it is only the use of 'must V' in the subjunctive way which permits the necessity to be viewed under different aspects. We can say that legally, morally, contractually or financially he had to V, but it makes no sense to say that legally, morally or financially he must have Ved. 'He must have Ved' is a conclusion which can be reached only on factual grounds even when it concerns legal, moral or financial matters. Though we can speak of the evidence or the reasons for thinking that he had to V or that he must have Ved, we can speak only of the reasons why he had to V and not of the reasons why he must have Ved. If A had to V, then it *was* necessary for A to V or necessary that he *should* V, whereas if A must have Ved, then it *is* necessary for it to be the case that he *did* V. Similarly, if he will have to V, it *will* be necessary *to* V (or that he should V); whereas if he must V, then it *is* necessary for it to be the case that he *will* V. 'He had to V' or 'He will have to V' reports the necessity which exists at the time of the Ving because of a factor or factors then operative, whereas 'He must have Ved' or 'He must V' reports the timeless necessity which holds between some fact or facts and the fact that he Ved or will V. In the former the circumstances allow only one possibility to be taken; in the latter they allow only one possibility to be the case. This subjunctive-governing use of 'must'—and, therefore, of the past form 'had to' instead of 'must have'—is not, as has sometimes been alleged,[1*] a distinguishing mark of human or even animal agency. It makes perfectly good sense to say such things as 'He had to die sometime', 'One has to be over 70' and even 'It had to happen'.

Though these various differences between 'must' used with the subjunctive and 'must' used with the indicative are exactly analogous to the differences between 'possible that' with the subjunctive (that is, the possibility expressed by 'can') and 'possible that' with the indicative (that is, the possibility expressed by 'may'), it would be a mistake to suppose that the two words are co-ordinate and that 'must' is sometimes contrasted with 'can' and sometimes with 'may'. For, the difference between 'must' with the subjunctive and 'must' with the indicative is not equivalent to the difference between necessity and certainty, but to the

* References to this chapter begin on p. 101.

difference between the necessity for something to V and the necessity for it to be the case that something Vs—or, in other words, between the necessity that something should V and the necessity that it should be that something does V. Similarly, the difference between 'must' with the subjunctive and 'must' with the indicative is not contrasted with the difference between 'possible that' with the subjunctive and 'possible that' with the indicative, but with the difference between its being possible for something to V and its being possible for it to be the case that something does V.

Any problem of what must be arises in a situation identified in terms of a set of circumstances, a requirement and a set of alternatives all viewed under a particular aspect. What must be is what amongst the alternatives is necessary in order that the requirement be met; what is the *only* alternative possible given the circumstances and the requirement. In the indicative use of 'A must V', what makes it necessary for it to be the case that A Vs will be whatever circumstances make it not logically possible, whether deductively or inductively, to have those circumstances and it not to be the case that A Vs. For instance, 'If the escaped prisoner went into the building a few moments ago and has not left, he must still be inside' or 'If the number was divisible by 18, it must divisible by 3'. In the subjunctive use of 'A must V', on the other hand, the aspects under which the necessity can be considered are legion. For example, anyone who drives a car must physically not be completely paralysed, must legally be insured for third party risks and must morally be considerate to other road users. Anyone in authority must psychologically have certain abilities, legally have certain powers and morally deal fairly and impartially. It is because of this variety of aspects under which something may be necessary, in the subjunctive use of 'must V' as contrasted with the indicative use, that 'must' and 'have to' are in the former, but not in the latter, often qualified by such adverbs as 'legally', 'economically', 'morally', etc.

The relation of the alternatives which must be taken to the circumstances and requirement is expressed in such phrases as 'in order to F' or 'for X', 'because of Z', 'in Y', e.g. 'In order to get home in time', 'for peace and quiet', 'because of the approaching danger', 'in all honesty'. Given the circumstances and the requirement, reasons can always be demanded for the contention that what must be is so-and-so. Such reasons will consist in a claim that certain features either of the circumstances or of the requirement or of one of the alternatives make this alternative what is in the circumstances necessary in order to meet the requirement. Each kind of reason is a complement to the other kinds. For example, you must work harder because there is not much time left and/or because you want to get a First and/or because only harder work will make a First possible.

Although a reason is always relevant to the contention that something must be, it is a mistake[2] to suppose that the notion of *must* has to be analysed in terms of a particular kind of *reason for*.

The difference between the indicative and the subjunctive uses of 'must V' is due to a syntactic difference in the way in which 'V' occurs in the two and not to any difference in the meaning of 'must'. Hence, any suggestion that 'must' can be defined in terms of the type of judgement in which it appears subjunctively with 'V' is mistaken. The fact that 'A must V' in its subjunctive use can be used to make 'practical' judgements about what to do is one consequence of the meaning of 'must' and not part of the explanation of its meaning nor of what it is to be a practical judgement.

NOTES

1. E.g. D. G. Brown, pp. 70–80.
2. E.g. Edgley, pp. 132 ff.; D. G. Brown, § 3.9.

Chapter Eight

Need

As part of any investigation of the modals, it is important to distinguish the notion expressed by 'need' from other species of necessity and particularly from the notion expressed by 'must'. It is, as we shall see, also important to distinguish the notion of *need* from two others with which it has often been confused, namely that of *lack* and that of *want*. A confusion of *need* with *lack* is inherent in much work in psychology, while a confusion of *need* with *want* pervades not only psychology, but also political and educational thought.

(A) NEED AND MUST

Although to say that A 'needs' to V means, as we shall see, that it is necessary for A to V, it is not the same as saying that A 'must' V. The former implies but is not implied by the latter. If A needs to V, then he must or has to V, but it can be true that A must or has to V without its being true that he needs to V. There are two reasons for this. First, as we saw, the notion expressed by 'must' can be used either with the *indicative* to signify that it is necessary for it to be that something *is* (was or will be) so, as in 'If he is already drawing an old-age pension, then he must be at least 70', or with the *subjunctive* to signify that it is necessary for it to be that something *should be* so, as in 'If he is to qualify for an old-age pension, he must be at least 70'. 'Need', however, cannot be used in the indicative way. Hence, though we can say 'If he is to qualify for an old-age pension, he needs to be at least 70 now', we cannot say 'If he is already drawing an old-age pension, he needs to be at least 70 now'. To say that the play must be about to begin—e.g. because it is now nearly

8 o'clock—is not to say that it needs to be about to begin—e.g. if it is to be finished by 10 o'clock. To say that I must have dropped my glove somewhere—e.g. because it is not in my pocket now—is not to say that I need to have dropped it somewhere—e.g. in order to leave a clue.

Secondly, even where the notion expressed by 'must' is used subjunctively to signify what it is necessary for it to be that there should be, a further distinction can, as we also saw, be made between its retrospective and its prospective use. That is, between what is necessary because it is *necessitated by* something and what is necessary because it is *necessary for* something. Only the latter is a need. Thus, a necessary result or consequence is one which has to be, but not necessarily one that is needed; whereas a necessary means is one which both has to be and is needed. To say 'If a gas is touched by a spark, it must explode' is different from saying 'If a gas is to explode, it must be touched by a spark'; only the latter signifies a need. A law which says that every citizen *must* have an identity card is making a different type of regulation from one which says that everyone going abroad *needs* to have—and, therefore of course, *must* have—a passport. There could well be a reason why I have to have, that is, am necessitated to have, something which I do not need to have. There is a subtle, but important, difference between 'I had to take this' and 'I needed to take this', between 'You must come and visit us' and 'You need to come and visit us'. The necessity expressed by 'need' is, therefore, a prospective necessity, that is a necessity *for* something. If A *needs* to V, there must be something he has to V for (in order to, etc.); whereas he can *have* to V without having to V for anything.

(B) NEED

Let us, therefore, first consider a few of the logical characteristics of the notion of *need*.

The basic grammatical[1]* constructions are 'A needs X' and 'A needs to V'. Here there are no logical restrictions on the sort of

* References to this chapter begin on p. 121.

thing, whether animate or inanimate, which can need; nor on the sort of thing, whether X or to V, which can be needed; though, of course, there are many sorts of things which a particular type of thing cannot need and many types of things which cannot need a particular sort of thing. Thus, a man can need a drink or to know what to do, an engine can need more oil or to be overhauled and a triangle can need at least two acute angles or to be revolved on its axis; though a man cannot need to be overhauled or a triangle need a drink. But there is no reason for supposing that the meaning of 'to need' depends on its subject and, therefore, that human beings and animals need anything in a different sense from that in which an industry or an engine needs something.

Whenever A needs X or needs to V, there is also a, or the, need for A to have X or to V; and *vice versa.* But such a need is not necessarily A's need; and, therefore, although if A's need is for X or to V, then A needs X or to V, the converse is not necessarily true. In other words, to have a need implies to need, whereas to need does not necessarily imply to have a need, though it does imply that there is a need. Thus, a student whose greatest need is more money needs more money just as an industry whose greatest need is more capital needs more capital; but a student who needs to work harder to pass an examination does not have a need for harder work—though there is a need for him to work harder—any more than an industry which needs to be liquidated has a need of liquidation. A valid syllogism needs at least one universal premiss and my dining-room curtains need to be lengthened, but neither valid syllogisms nor my curtains have any needs. Both A's *needing* something and *what* A needs are called 'A's need',[2] just as both A's want of, hope for or pleasure in something and what A wants, hopes for or takes pleasure in are called his want, hope or pleasure. A's needs, therefore, can be either food and water or a need for food and a need for water. It is A's need, in the sense of his needing, which he has, feels or can lose, which is met or overlooked, which is great or small, conflicting, pressing or immediate, just as it is his want, in the sense of his wanting, which he has or is seized by, which is insatiable or unconscious. When Sir Philip Sydney replied 'Your need is greater than mine', he meant that his companion needed the water more than he did. To feel a need for water is not to feel the water.

To say that A needs X or needs to V is to say that X or to V, that is, what A needs, is something which is necessary for A. A's need of something is its being thus necessary for him and what A needs is what is thus necessary. 'Need' indicates a relation, namely that of a certain kind of necessity, between one of a number of alternatives, which is said to be what is needed, and a situation which consists of a set of circumstances and an end-state. A cannot in these circumstances reach the end-state without Ving or without X. The end-state—which can be discovered or stipulated—is what A needs to V *for*, e.g. for success in life, for peace and quiet, for reduction of engine wear or—and this is very common—to be an A; and the circumstances are what A needs to V *because of*, e.g. because of the ravages of war, the heavy unemployment or the friction in the cylinders. Because the end-state can be of many kinds, to say that A needs to V is elliptical for saying that A needs to V in order to F. Furthermore, what is needed can be extrinsic to the end-state, as when a man needs a lot of money to be well-dressed or a number needs to be small to be remembered, or it can be intrinsic, as when a man needs a certain style of clothes to be well-dressed or a number needs to be unfactorizable to be a prime. The need can be further qualified in regard to the aspect under which it is related to the end-state, as 'legally', 'logically', 'morally', 'psychologically', 'practically', 'ideally', 'in theory', 'according to the regulations', etc.[3] Hence, A can morally need to V, though there is no legal necessity for it; or ideally need to V without practically needing to. But whether needs are chosen—e.g. because I want to attain some end—or are imposed—e.g. by physical or legal circumstances—whether they are part of the course of life—e.g. biological or psychological—or merely adventitious—e.g. for a momentary end—makes no difference to the logic of *need*. Hence, it is a mistake[4] to suppose that the existence of such various categories of need forms part of the structure of the concept of *need*.

Given the circumstances and the end-state, reasons can always be demanded for the contention that what A needs is so-and-so. Such reasons will be given by statements about how the circumstances, the end-state or some features of the alternatives make this alternative what is needed in the circumstances to lead to this end-state. Each kind of reason is complementary to the other kinds. Thus, it may be that A needs to work harder either because he has been slacking or in order to get a First or because it is not possible to get a First without hard work. He may need a passport either because he has lost his or because he is going abroad or because a passport is a necessary requirement for foreign travel. But though a reason is in this way always relevant to the contention that A needs so-and-so, it would be a mistake[5] to suppose that the notion of *need*, or any other modal, can be analysed in terms of a reason for so-and-so.

Commonly, the context makes clear what are the circumstances, the end-state and the aspect in respect to which A needs to V. But a failure to notice the elliptical nature of statements about what A needs leads to arguments at cross purposes, for it can easily be the case that although legally A needs to V, he does not physically or morally need to, or, although he needs to V in order to F he does not need to V in order to G. When, therefore, someone disputes that A needs to V, he may be disputing this either because he does not think that given the circumstances and the end-state it follows that to V is something the absence of which would be the lack of something necessary or because he wants the question whether A needs to V considered in relation to a different end. My daughter's claim that she needs another pair of shoes is made in the light of her wish to keep up with the latest fashions; while my insistence that she does not need them appeals to the ordinary requirements of daily use. There is no non-relative answer to the question 'Does she need them?', though one may legitimately suppose that the need to V in order to F is more important than the need to V in order to G because to F is more important than to G.

This relative nature of a need is, however, misinterpreted in the current philosophical view that 'the term "need" is mainly normative'.[6] Whether A needs to V or not depends solely on whether or not Ving is in the circumstances the only way to reach the end-state, and whether there is such an end-state. If I am going abroad, I need a passport; if the ball-bearing is to last a year, it needs oil; if you are no longer Dean, you don't need those files. It is just as much a question of fact whether there is such an end-state as whether such an alternative is necessary to reach it. The question whether there should or need be such an end-state is a different question and quite independent of the logic of *need*. 'Does A need X?' is an elliptical, not a normative, question and, hence, one that can be answered only when expanded, e.g., into 'Does A need two A-levels in order to get into (since he is going to or wants to go to) an English university?' but not into any question about whether getting into (going to or wanting to go to) an English university is valuable or not. Nor is the logic of *need* altered by the fact that the end-state because of which A needs to V can express an evaluation, as when a man needs either a lot of money or a certain style of clothes in order to be well-dressed as contrasted with being dressed in the height of fashion. Even if the statement 'A needs X in order to be a good (true, proper) A' is normative, 'A needs X in order to be an A' is not. Nor does occurrence in a normative statement necessarily make a term normative. 'X is conducive to, a sufficient condition of, or an expression of, goodness in A' can be normative without 'conducive to', 'sufficient condition of' or 'expression of' being normative terms.

Often, what A needs will be something he or it actually *lacks*, as when a man needs more exercise or an engine needs a new piston; though what he needs and lacks may itself be something which diminishes rather than adds to him, as when a man needs a hair-cut or needs to be taken down a peg. And, perhaps, the 'needy' and those 'in need' do lack something. But A can also need what he or it already has, as when a man needs all the money he has got or an engine needs all the attention it gets. The little girl who tells me that she needs 10p for the bus probably does not have the 10p; but when she tells me that she needs the 10p in her pocket for sweets, she certainly has the 10p. Therefore, the fact that she needs 10p for so-and-so does not imply either that she has 10p or that she has not. Hence, it is a mistake to suppose, as is commonly done,[7] that to need something is to lack it. If to need something were to lack it, then to complain 'You are never here when I need you' would be to rail against a logical necessity. Surely, I could properly feel insulted if you were to conclude from my remark that on such-and-such an occasion I needed all my wits about me that, therefore, I was on that occasion somewhat witless. We can say 'I have all the things I need' and not only 'I have all the things I needed'. Both what is not absent and what is absent can be something whose absence would be the lack of something necessary. Nor, conversely, does lacking something imply needing it.[8] There are lots of things, like a taste for blood sports, which I neither have nor need. My car can lack a sunshine roof without needing it and, contrapositively, the fact that I do not need a visa to travel to Ireland does not imply that I do not lack such a visa. If to lack something were to need it, it would be self-defeating to say 'You can have this, I don't need it any longer'.

(c) NEEDS AND WANTS

Needs are often and importantly confused with *wants* and *needing* confused with *wanting*, partly because of linguistic ambiguities, in English at least, but mainly because of logical similarities.

Linguistically, the word 'want' is used in at least two other ways additional to its use to express a want. First, there is a dialect use of the word 'want',[9] especially common in the north of England, which replaces the word 'need' both for animate and inanimate subjects. Thus, a driver may be told that he 'wants' to turn second left in order to reach the main road, a plant is said to 'want' water, a door which needs to be closed is said 'to want closing' and a man who needs to be watched is said 'to want watching'. But this dialect use is still rather confined. It is still not ordinarily said, e.g., that the denominator of a proper fraction 'wants' to be greater than the numerator. Secondly, there is a colloquial use of the word 'want' in which it means the same as the word 'lack', e.g. 'for want of a nail', 'to be found wanting', 'to want courage'. The occurrence of this use may also lead some people to think that 'wanting' is the same as 'needing' because they suppose, wrongly as we saw, that 'needing' is the same as 'lacking'.

Logically, what I want—in the common use of 'want'—is like what I need in several respects. First, what I want, like what I need, is either something X or to V; though, as we shall see, there are severe restrictions both on what can be wanted and on what can want it which are absent in the case of *need.* Secondly, one cannot have a want or a need without its being a want for or a need of something.

Thirdly, one can want or need, as one can look for, something, e.g. an inexhaustible supply of money, which does not exist. Furthermore, to say that A wants, needs or is looking for *an* X does not imply that there is a particular X which he wants, needs or is looking for. Hence, though A may need or want—as he may look for—an aspirin, it would be absurd to demand to know which of several (equally suitable) aspirins he wants or needs—or is looking for; an absurdity which a colleague dubbed 'Buridan's aspirin'. When, therefore, someone who, wanting or needing or looking for a new suit, announces a discovery by saying '*That* is just what I wanted (needed or was looking for)', he means 'That, namely a suit of that kind' and not 'That, namely that particular suit'.

Fourthly, there is no necessary suggestion[10] in 'A wants X or wants to V'—any more than in 'A needs X or needs to V'—that A must either lack X or not already be Ving. I can want what I have and want to do what I am doing. Nor need 'I want to have what I have or to do what I am doing' mean, as Aquinas suggested,[11] 'I want to keep on having or doing what I have been having or doing'. For, if I wanted to kill someone whom I did kill, it does not follow that I did not want to kill him while I was killing him nor that I wanted, *per impossibile*, to keep on killing him. Furthermore, a man can now have not merely what he want*ed*, but also what he want*s*. The supposition that one can only want what one has not got partly depends on a confusion of wanting with lacking and partly on the assumption[12] that to want is to want to get. But I can want to get, to hold, to lose or to forget. This common philosophical linking of wanting with getting is paralleled by the equally common linking of wanting with attaining some end. Philosophers have over-concentrated on examples like 'I want my car today, so I'll go to the car-park' to the exclusion of examples like 'I want my car today, so I cannot lend it to you'.

Despite these similarities, the notions of *need* and *want* are very different,[13] both grammatically[14] and logically. Here I shall discuss only the logical differences.

Want, unlike *need*, carries no necessary reference to an end-state in virtue of which something is wanted. Hence, there is nothing logically elliptical in a remark that 'A wants so-and-so'. We cannot dispute whether someone wants a television set or a second car, as we can whether he needs them, by querying whether there is an implied end-state or whether what is wanted would attain this. Further, want, unlike need, is not relative to an aspect. We cannot speak of what someone morally, legally, economically, theoretically, in all fairness, etc. wants, as we can of what he needs.

Hence, the idea of a reason is differently related to *needs* and *wants*. Someone can want to V either for some reason or just because he wants to V; but one cannot need to V just because one needs to V. There has to be an explanation of why he needs to V. One explains why something is needed by showing that it is necessary; but one explains why something is wanted by showing that it is thought desirable either in itself or as a means to something else. To say you do something because you need to suggests a constraint, whether physical, legal, moral, or whatever; whereas to say you do something because you want to expresses your independence of any such constraint. To say that one wants to V is, in certain circumstances, to explain one's actions; but to say that one needs to V is, in comparable circumstances, to try to justify them. Furthermore, where one can both want to V and need to V in order to F, e.g. to want to or need to move one's bishop in order to check one's opponent, to F in the latter is the result for which Ving is a necessary condition, whereas to F in the former is one's reason for desiring to V. To tell someone that he does not need so-and-so for the purpose for which he said he did is to correct his general knowledge; to tell him that he does not want it for that purpose is to question either his sincerity or his self-knowledge. One can say both that A wanted and needed a British passport in order to get the protection of the British consul; but, whereas it makes sense to say that in order to get a British passport A needs to have been born in the United Kingdom, it is nonsense to say that in order to get a British passport A wants to have been born in the United Kingdom.

It is, indeed, debatable[15] whether one can want—as contrasted with wishing—something to have been; but there is no difficulty in needing this, e.g. in needing to have been born in the United Kingdom. There is, further, an oddness, though not an impossibility, about wanting—as opposed to wishing for—what one knows to be logically impossible; but something, e.g. an argument, can properly be said to need what is known to be logically impossible in order to be what it, therefore, cannot be, e.g. consistent or valid.

An extremely important difference between *want* and *need* is revealed by the fact that—apart from the dialect use of 'want'—only animate creatures can be said to want, but anything can need. Adults, children, animals, the country's economy, the coal industry, an internal combustion engine and a proper fraction may need different things, but they all need them in the same sense of 'need'.[16] But we cannot talk of anything other than the animate or, perhaps, even the intelligent wanting something.

One consequence of this is that verbs which can take only inanimate or impersonal subjects cannot follow 'want' as they can 'need'. It makes no sense to say that A wanted to happen, to occur, to be proved or to be a prime. On the other hand, epithets assessing the character of an agent who wants something are not applicable to the same agent when he needs it. It can be kind, good, generous, selfish, greedy, illogical, sensible or unreasonable of someone to want to V, but not to need to V. An agent can ardently, eagerly, impetuously, rashly or foolishly want to V, but not in these ways need to V. One can control or indulge one's wants, but not one's needs. We praise or blame a person because of what he wants—or because of what he does—but not because of what he needs.

Because wants are confined to animate subjects, there is a connection between wants and beliefs which is lacking between needs and beliefs. What I want to do often, and perhaps usually, depends on what I believe. I want to do so-and-so because I believe it has such-and-such characteristics just as I fear, dislike or admire so-and-so for these sorts of reasons. Usually, perhaps, these characteristics must be ones which I think desirable.[17] But what I need is independent of what I believe. Whether I need so-and-so partly depends on the characteristics of so-and-so, whether or not these are desirable, but not at all on what I believe to be or believe of these characteristics. Any such belief could at most only make me think I needed it. 'I wouldn't want to do it, if I didn't think it was necessary (or desirable)' expresses a common and reasonable position; but to say 'I wouldn't need to do it, if I didn't think it was necessary (or desirable)' is only to fool one's self.

For the same reasons, what someone avows he wants is good—some might think the ultimate—evidence for what he wants; but what he avows he needs plays little part in deciding on his needs.

One can want something under one description, but not under another. To want to kill the man who is blocking your escape does not imply wanting to kill your own son, even though it is your son who is blocking your escape. What one needs, on the other hand, one needs whatever its description. If, in order to escape, one needs to kill the man blocking one's way and the man is one's own son, then one needs to kill one's own son. The reason for this difference is that if I either want or need—or seek or hope—to do something which, in the circumstances, I can do *only* by doing something else, then I need to do this other thing, whether I want (seek or hope) to do it or not. The world is full of necessary evils and regrettable necessities.

Incidentally, it is because the identification of what one wants is linked to the description under which one wants it that great care has to be taken in attributing particular wants to animals. Suppose we place a large, round, red object in front of an animal. Does the animal want to attack or flee from the red object, a round object, a large object or some such set of these? No such problem arises in trying to discover the needs of animals.

We saw earlier that there are very important differences between the way in which reasons are related to wants and the way in which they are related to needs. A further difference is this. There can be reasons both why I want and why I need so-and-so; but only for wanting so-and-so can there be such a thing as *my* reasons. 'I have my reasons for wanting him out of the way (or for wanting a change of government)' makes sense; but not 'I have my reasons for needing him out of the way (or for needing a change of government)'. Moreover, since something can be my reason for wanting to do so-and-so—or for doing so-and-so—it can be a good or a bad reason for it. But there can be no good or bad reasons for needing to do so-and-so; there either is or there is not a reason why I need to do so-and-so. In this respect, 'wanting' is analogous to 'feeling annoyed', while 'needing' is analogous to 'feeling an itch'. There can *be a reason* both for my feeling annoyed and for my feeling an itch; but *I can have a reason* only for the former.

The great difference between need and want is clearly revealed, therefore, in the distinction between what can characterize both an animate and an inanimate subject and what is limited to an animate subject. Nevertheless, even when needs and wants characterize the same, and therefore an animate, subject, they are still different. It is easy to show this.

First, even with an animate subject, the idea of a want seems more confined than that of a need. Where there is no suggestion that anyone, whether the subject or another, has logically any power to V, the notion of wanting to V goes less easily. Thus, we cannot want to expect, prefer, fancy, imagine, regret, envy, dread or mind anything, but it makes sense to say that one needs or does not need to do any of these. A special case of this is the past. We saw earlier that 'A needs to have been born a British subject' makes sense, but not 'A wants to have been born'—as contrasted with 'A wishes that he had been born'—'a British subject'. Another example of lack of power is where the occurrence of a certain result is a reception rather than an accomplishment and, therefore, not something in whose occurrence anyone could have had any hand. One cannot want, as one can need, to recognize, realize, notice, be aware or beware of, infer or deduce, something.

Secondly, even where the notions of *want* and *need* are both applicable there is no mutual implication between them. What A wants is not necessarily what he needs; nor *vice versa.* Someone can want more money, though he does not need it; and he can need a good hiding, though he does not want it. Reluctant people often do not want to do what they need to do, while generous people often want to do what they do not need to do. Telling a person that he does not need to believe you is quite different from telling him that he does not want to believe you. The instruction 'Do whatever you need to do' is quite different from 'Do whatever you want to do'. Any action taken on the first instruction would require a quite different and much more objective defence than any taken on the second. The second instruction gives a *carte blanche* which the first does not. Knowing full well that I need to do so-and-so, I can still ask myself whether or not I want to do it. Contrariwise, 'You don't need to, if you don't want to' is not a tautology. It is a 'degenerate' exception to this lack of implication between wanting and needing that 'A wants X' implies that A needs X in order to satisfy his want; for the former lack of implication is applicable only to cases where what A wants X and needs X for are the same.

Thirdly, it is plausible to hold that what a man really wants to do, he will, other things being equal, actually do; but there is no plausibility in the suggestion that what he really needs to do, he will, other things being equal, actually do. Trying to get may be 'a primitive sign of wanting';[18] but it is irrelevant to needing.

Fourthly, the psychological element which is central to *want* is absent from *need* even when what is in question is a need of a human being. Hence, as we saw, (a) one can want something because one believes that it possesses certain characteristics, whether it does or not, but one can need something only because it actually does possess certain characteristics. (b) To explain my action by reference to my wants may be to explain them by *my* reasons; to explain then by reference to my needs—as distinct from my realization of my needs—can only be by *the* reasons.

Further, (c) though one can feel a need, as one can feel a desire —though perhaps not a want—for something such as food, warmth, love or comfort; one can feel a need, but not a desire, which one does not really have. Conversely, one can have a need, but not a desire (or want?), which one does not feel. Moreover, most things for which one has a need, unlike those for which one has a desire (or want?), are not things one can feel a need for, e.g. more salt in one's diet, a more frequent bus service, though they may be things for which one can feel that there is a need. (d) *Emotion* logically involves *want* and *inclination*; e.g. to fear implies to want and feel inclined to take avoiding action, to be angry implies to want and feel inclined to retaliate. But *need* plays no part in the analysis of *emotion*, even though there may be a psychological connection of one's emotions, e.g. love or fear, with various bodily and mental needs.

(e) It is arguable[19] that, in some sense, we always know what we want, but not at all plausible to suppose that we always know what we need. We have no hesitation in allowing that experts know better than ourselves what we need, but are reluctant to admit that anyone knows better than ourselves what we want. It is at least doubtful whether there can be something which I clearly and obviously want, but don't know this; but it is not at all uncommon for there to be something which I clearly and obviously need, but don't know this. '*How* do you know that you want to go?' is queer; whereas '*How* do you know that you need to go?' is perfectly normal. 'He does not know what he wants' may be used mainly of someone who cannot make up, or is constantly changing, his mind; but 'He does not know what he needs' is used simply to state his ignorance of some necessity. A psychoanalyst may agree that to call something an unconscious want, or desire, is to say that it is something which the patient can be brought, with suitable treatment or suggestion, to admit to have been a want or desire; but there is no reason to insist that something can be a need of which one was unaware only if one can be brought to acknowledge it was a need.

(f) Because a man usually knows what he wants and usually bases his conduct on what he wants, we can often guess or discover what he wants from observing his behaviour. Indeed, one might reasonably hold that to discover what a baby or an animal wants is to interpret his behaviour as a manifestation or expression of his wants rather than as an effect of them. But there is no such close connection between what someone needs and the behaviour he exhibits. His behaviour can, of course, sometimes give a clue to his needs, but it is not a manifestation or expression of them. What shows that a dog needs to go for a walk or that a baby needs his nappy changed is not necessarily what shows that the dog wants to go for a walk or that the baby wants his nappy changed. One's wants are connected with one's psychological reactions. If a man is not disappointed at not getting so-and-so, this is good evidence that he did not want it, but not that he did not need it. The role of reactions is quite different for needs, even psychological needs. The way in which biological or psychological reactions are related to needs is not logically different from the way in which, e.g., the reactions of an engine to certain tests show whether it needs more oil.

(g) The psychological element present in wanting, but absent in needing, is shown also in the confinement of a want to what the agent is capable of understanding, and, therefore, of expressing. One cannot want something without having the idea of that which one wants; but one can need something without any idea of what one needs. There is no doubt that a dog may need—and his behaviour may show that he needs—to be injected against distemper; but he, as contrasted with a human who understood the idea of inoculation against disease, could not properly be said to want, even in a behaviouristic way, to be injected against distemper. We can say that a dog wants to go for a walk now, but not that he wants to go for a walk next Good Friday.[20] There is no similar restriction on what he needs to do.

(D) PSYCHOLOGY, POLITICS AND EDUCATION

Having explored the relations between the notions of *need* and *want*, let us draw some morals in three fields where, I suggested, the notions frequently occur, namely in the philosophy of psychology, in political philosophy, and in philosophy of education.

(1) The notion of *need*, especially as applied to psychological and biological needs like the need for food and drink, comfort and affection, appeared prominently in the psychology of the immediate pre- and post-war periods (1930–60) in attempts to explain human and animal behaviour. From there it crept into educational psychology. In much of the literature,[21] the notion seems to have been partly confused with the notion of *want*, partly confused with that of *lack* and partly treated on a false analogy with both.

But psychological needs can no more be identified with wants than can any other kind of needs.[22] For to call a need psychological or biological, like calling it economic or legal, is only to say that the aspect under which it is considered is of a certain kind. Needing food and water is not the same as wanting them and, therefore, not the same as being hungry and thirsty.[23] In abnormal cases people can want to drink even when they have no further need of water and can be without any desire to drink even when they still need water. Plants need water as much as animals, but it is only in an anthropomorphic sense that they can be said to be thirsty, that is, to want to drink. This is not, of course, to deny that there may be contingent connections between wants and certain needs. Perhaps, it is because we need water that we want to drink. But these connections can only be contingent. A dog which needs to be inoculated cannot, as we saw, want to be inoculated; though some of the things it wants, e.g. to scratch or to run, may be connected in biological ways with its need to be inoculated.

Since, as we saw earlier, the kind of explanation that a need supplies for something, e.g., needing water in order to keep alive, is not the same kind as that which a want supplies, to replace explanations of human and animal behaviour in terms of wants by explanations in terms of needs is to give quite a different kind of explanation.[24]

Whereas some psychologists seem to have assimilated needs to wants and both to motives or drives, others, in distinguishing wants and drives from their sources, cast needs in the role of the source of drives. In doing so, they usually identified needs with lacks or deficits.[25] 'Need-reduction' was introduced as a synonym for 'filling up a lack'. We have already seen that whether or not a lack of something gives rise to a need for it and whether either of these makes somebody want it are contingent empirical questions to which the answer is sometimes 'yes' and often 'no'. They certainly are not necessarily affirmative as they would be if needs were the same as either lacks or wants.

Both the assimilation of needs to wants and their assimilation to lacks are probably partly responsible for the common talk about 'states of need'. Thus, a typical psychological definition of a 'need state' is 'a condition of deficit or excess which causes the organism to depart from the equilibrium known as homeostasis'.[26] Such a deficit or lack is often inferred from a long period of deprivation or restriction, e.g. of food, drink, sex or movement. But, whereas a want might possibly be called a 'psychological state', a need, even a psychological need, is not a psychological state. A need of food or security is no more a psychological state than a need of a passport or a motor car is some kind of state. Nor is having a need the same as being in a state of tension. To meet a need is not necessarily to reduce a tension. 'His need of money drove him to a life of crime' is logically different from 'His depression drove him to drink'.

If an organism needs something for survival, e.g. a plant needs light or an animal needs water, then, no doubt, there will be some biological or psychological conditions which generate the need; and others which help to meet it. But the need for light or water must not be identified with any of these conditions. It is perfectly proper to say 'If food is withheld (from an organism) for a sufficient period of time, the chemical structure of the body is altered, and a need for food is said to exist because the ingestion of food is the necessary condition for the restoration of the original state.'[27] But we should not then reify a need as a psychological, physiological or chemical state of the organism of which this is true any more than we identify the decrepit state of a house which badly needs repair with an architectural state of need. Nor should one identify the organism's need of food with a food-drive any more than the house's need of repair with its liability to fall down. The error here is similar to that, also common in these psychologists,[28] of assimilating a habit to the mechanism which accounts for the habit. Psychologists[29] have wrongly inferred from the fact that A has a need for X that a need is a psychological or physiological state in which A is when he has the need.

(2) Of the various contexts in political philosophy in which the notions of *need* and *want* occur, I shall mention only that of public and private interests.

Political philosophers are prone to connect people's interests either with what they want or with what they ought to want or with what will help them to get what they want.[30] But this is a mistake due to confusing the different ideas of *A's being interested in X* and *X's being in or to A's interest.* What is to the best of one's interests may not be what one is interested in; and *vice versa.* Being in a particular matter an interested party does not prevent one from being bored. What the public is interested in, e.g. gossip about its neighbours, need not be in the public interest, e.g. in the interests of health and wealth. Similarly, one's interests can be those things in which one is interested, e.g. music, books or stamp collecting, or those things which are to one's interest, e.g. health or the provision of good roads. If I consult your interests when I am engaged in forming a policy, it is the latter, not the former, which I take into consideration.

What one is interested in is a sub-class of what one wants, namely, what one wants—and feels inclined—to give one's attention to; whereas one's interests are at most a sub-class of what one needs, namely, what one needs for one's well-being. Hence, to study someone's interests is to have regard to some of his needs rather than to some of his wants. And just as we praise or blame a person for having certain sorts of wants, but not for having certain sorts of needs, we praise or blame him for being interested in so-and-so, but not because such-and-such is in his interests. When that whose interests we consider is not a person, but an animal or thing, e.g. in the interests of wildlife, of our country, of the economy, of peace, harmony, safety or truth, what we consult is what it needs and not, *per impossibile*, what it wants.

Deciding in favour of public or private interests is deciding in favour of needs, not wants. What is in the public interest is not necessarily what the public wants. When an Act of Parliament (e.g. The Restrictive Trade Practices Act 1956) declares something to be or not to be contrary to the public interest, it does so on the grounds of what is necessary or not necessary for the attainment of certain objectives. The idea of a (or the) general interest arises from the fact that people living together develop common needs for such things as law and order, defence and education.

'Discovering someone's interests' is, unfortunately, ambiguous between 'discovering what he is interested in' and 'discovering what is in his interest'. To discover the former is to discover what he wants to give his attention to, whereas to discover the latter is to discover what he needs for his well-being. Just as someone may not know what he needs, so he may not know what is in his interests; though he will as surely know what he is interested in as he knows what he wants. It is only if one wrongly analyses *being in one's interests* in terms of wants, that one has to hold that since people often do not know what is in their best interests, they do not really know what they want. Parents usually ask their children what they are interested in, but often decide for themselves what is in their children's best interests. Whitehall may know better than we do what is in our interests. There is, therefore, no inconsistency[31] in the political suggestion that someone's true interests may lie elsewhere than his own wants and inclinations. It is because modern industry needs a large body of consumers that joining the Common Market may be in our interests even if we don't want to join.

I don't wish to overstress the relation of needs to interests, but rather to argue against the common linking, in political philosophy, of interests to wants. What is in someone's interests may be more a question of what is *best* for him than what is *necessary* for him; but what is best for him is more dependent on what he needs for his well-being than on what he wants to have. It is a worthy moral view that we should consider the needs of others and a worthy political view that we should consider the interests of others, when deciding what to do. But I can see no great virtue in the quite different view that in deciding what to do we should consider exclusively what others want and are interested in. The latter view gains much of its plausibility from assimilating needs to wants.[32]

(3) The final area in which I wish to consider the notions of *need* and *want* is that of the philosophy of education. Here we clearly have to distinguish a pupil's needs from his wants, what is in his interest from what he is interested in. Educationalists, but not analytic philosophers, can and do dispute which of these ought to be taken into account in the classroom. Dewey had his followers and his opponents. It is a psychological, not a logical, assumption that pandering to a child's interest is catering for his interests, and it is a statement of principle, not of logic, that a school should meet some of the non-educational needs of its pupils.

But certain conclusions logically follow from the fact, which we have investigated, that needs differ from wants; and these are not matters of educational values or of psychological expertise. For example, a child may be the best judge of his wants, but not necessarily of his needs. A teacher can appeal to or catch his interest, but not his interests. Though teachers sensibly take advantage of the fact that some of what a child wants to learn and some of what he needs to learn happily coincide, questions of motivation are linked to the pupil's wants and questions of education to his needs. That educational institutions should meet the educational needs of their pupils is a tautology; that they should meet their educational wants or wishes is a debatable piece of policy as is the view that they should meet some of their non-educational needs.

Recent writers[33] have emphasized these differences so clearly that there are only two objections I wish to make. The first and minor one is to regret the still common assumption that what we need must necessarily be something we lack.[34] But all the children in the class need affection, both those who get it and those who don't. My major disagreement is with the fashionable tendency to assert that statements about needs are necessarily normative.[35] As we saw earlier, they are certainly elliptical; that is, if something is needed, it is needed for . . . or in order to . . . or because of . . . or . . . or whatever; that is, with reference to some end-state. Furthermore, needs can be logical, legal, moral, economic, psychological, or whatever. But it is the *existence* of the end-state, not its *value* or *desirability*, that makes something needed. A batsman on 99 needs one more run for his century whatever one thinks about getting a century. Children need to understand addition in order to manage multiplication, whatever the value of either. Nor does the statement that A needs X become normative just because what A needs X for is itself something normative. 'A needs care and affection' has exactly the same (elliptical) role whether A needs it in order to become a good citizen or in order to become a wealthy citizen. The confusion between the *existence* and the *desirability* of an end-state has arisen from a confusion[36] between *specifying* the end-state in virtue of which the need arises and *justifying* it. One cannot decide whether A needs X unless one knows what he is alleged to need it for; but one does not have to pronounce on the merits of the latter.

NOTES

1. Grammatically, 'need' occurs, though in a rather unsystematic way, both as a (modal) auxiliary and as a full verb; cp. Palmer, pp. 37 ff. Sometimes, especially in the negative and interrogative forms, the 'to' is omitted, e.g. 'Need I go?' and 'You need no longer work'.
2. This twofold use accounts for the divergence among educational psychologists—noted by Komisar—on whether a need is the condition

of lacking an object or the object lacked. Both views, as we shall see, are mistaken.

3. Nor do these differences imply different uses of 'need'; *pace* Komisar, pp. 29–30.
4. Braybrooke, pp. 86–107. Although he lists many of the features of *need* which I have given, he says little about the nature of the notion itself.
5. E.g. D. G. Brown, pp. 70–80, 102; Edgley, pp. 133–45, 164–5.
6. E.g. Peters (1958), pp. 17–19; Sparshott, pp. 131–42; Benn and Peters, pp. 141 fl.; Hirst and Peters, pp. 32–6; Komisar; Dearden (1966), pp. 5–17; (1968), pp. 14 fl. Cp. Ehrman, p. 74, 'The predication is required by the speaker's or writer's view of some aspects of the world.' Nor is *need* at all like *ought*; *pace* Nielsen, pp. 186–206.
7. Sparshott, pp. 133–4; Hirst and Peters, p. 33; Komisar, p. 27 (contrast p. 31).
8. *Pace* Radford, p. 51.
9. Cp. Palmer, p. 167.
10. *Pace* Aquinas, *Summa Theologica*, Ia IIae, 30 2 ad 1. Grammarians sometimes say that 'want to' always refers to a future event, e.g. Palmer, p. 160.
11. *Op. cit.*; cp. Kenny, pp. 115–6, 119–20. On this see Radford and Hinton, pp. 78 fl.
12. E.g. Anscombe (1957), p. 67 fl.; Melden, pp. 119 fl.; Kenny, pp. 112–16. Contrast Matthews and Cohen, pp. 455–6.
13. *Pace* P. W. Taylor, pp. 106–11.
14. Grammatical differences (see Palmer, *passim*) show that 'need', but not 'want', is an auxiliary; e.g. we have 'needn't', but not 'wantn't', 'Need you ask?', but not 'Want you to ask?', 'need do', but not 'want do'. Though even 'need' differs from other auxiliaries in, e.g., avoiding repetition, such as 'I need to do it and so does he'.
15. Contrast Anscombe (1957), p. 69 with Kenny, p. 115.
16. *Pace* P. W. Taylor, p. 106; and Nielsen, p. 191.
17. Cp. Anscombe (1957), pp. 61 fl. and Gauthier, pp. 34 fl.
18. E.g. Anscombe (1957), p. 67.
19. Cp. McGuinness, pp. 305–20.
20. Cp. Wittgenstein (1953), § 650.
21. E.g. Hull; Maslow. For a survey of the literature, see Baumeister, Hawkins and Cromwell, p. 438–53. For criticisms, see Peters (1958) and Komisar.
22. *Pace* P. W. Taylor.
23. J. S. Brown distinguishes needs and drives.
24. Cp. the objections of Peters (1958) to various psychologists.
25. Cp. references in Baumeister *et al.* and Komisar.
26. Baumeister *et al.*
27. J. S. Brown, p. 67.
28. E.g. Hull.
29. E.g. Hull and Skinner, ch. 9.

30. E.g. Benn, pp. 123–40 (though contrast pp. 130–1); Miller, chs. 3–5; Barry and Rees, pp. 4 and 19–20.
31. *Pace* Miller.
32. E.g. in morals Gauthier, pp. 86 fl., 125.
33. E.g. Hirst and Peters; Dearden; Archambault, pp. 38–62; Wilson, chs. 1 and 2.
34. E.g. Komisar; Hirst and Peters.
35. E.g. Peters (1958); Benn and Peters; Komisar; Dearden; Wilson.
36. As in Wilson, ch. I; contrast P. W. Taylor, p. 111.

Chapter Nine

Obliged

A kind of necessity which frequently arises in human conduct is that signified by the notion of *being obliged.*[1]* Among the various ways in which a person's freedom of action may be diminished or restricted—e.g. he may be acted upon, he may be unable to help doing what he does, he may suffer from some disability or have a duty imposed on him—is that of being obliged to do something. Here I wish to examine this latter notion. In order to clarify it, I shall investigate the differences between 'being obliged by A to V' and 'being obliged to A for X', between 'being obliged to V' and 'having (or being under) an obligation to V' and between being physically obliged and being morally or legally obliged. In the next chapter, after an examination of the concept of what *ought* to be, I shall discuss the differences between 'I am obliged to V' and 'I ought to V'. An assimilation of these two is a hoary philosophical howler.

(A) 'OBLIGED BY' AND 'OBLIGED TO'

If I am obliged by somebody or something to do X, then he or it obliges me to do X. If, on the other hand, I am obliged to somebody, or perhaps something, for doing Y (or for Z), then he or it obliges me by doing Y (or with Z). For instance, in the first case, if I am obliged by the Vice-Chancellor, a university regulation or the illness of a colleague, to attend a meeting, then the Vice-Chancellor, a university regulation or the illness of a colleague, obliges me to attend a meeting. In the second case, if I am obliged to the desk clerk for (or for giving me) the required

* References to this chapter begin on p. 138.

information, then he obliges me with (or by giving me) the required information.

I am not obliged *to* what I am obliged *by*, nor obliged *by* what I am obliged *to*. The person, or thing, to whom I am obliged has helped me; the person, or thing, by whom I am obliged has not. The person to whom I am obliged is obliging, but not the person by whom I am obliged. An obliging person is necessarily helpful. I am obliged to someone for his doing of something, but obliged by someone only by his doing of something. I am obliged to someone because of what he has done for me, but obliged by him because of what he has done to me. I can be obliged to someone to a greater or lesser degree, but there are no degrees to which I can be obliged by someone. I can either be obliged to someone or be obliged by someone to do something, but I cannot be obliged to someone to do something nor can I be obliged by someone without being obliged by him to do something.

The one is, however, basically similar to the other. Whether I am obliged to someone for what he has done or obliged by him by what he has done, whether he simply obliges me or obliges me to do something, I am obliged by his doing of something. Further, though I cannot be obliged to myself for what I myself do, I can be obliged to do so-and-so by what I myself do, as when I am obliged to do X by my having promised or contracted to do it. Finally, when what obliges me to do something is not a person at all, but, e.g., a rule or a physical event or a psychological condition, then also there is something by which, though nothing to which, I am obliged. In every case of being obliged, therefore, whether I be obliged to someone or to do something, there is something, though not necessarily someone, by which I am obliged. That by which I am obliged may be called the 'obliging factor'.

Besides being obliged to or by, I can also feel obliged to or by someone or something or feel obliged to do so-and-so, irrespective of whether or not I am in fact obliged.

(B) 'BEING OBLIGED' AND 'HAVING (OR BEING UNDER) AN OBLIGATION'

I may have (or be under) an obligation to someone or to do something. An obligation may be assumed because of what I do, as when I make a promise, or incurred because of what someone has done for me, as when he helps me, or imposed because of what someone or something does to me, as when the law puts me under an obligation. I may acknowledge, fulfil or be released from such an obligation. I may enter into negotiations with or without obligation. I may receive something which carries or is free from obligation.

To have (or be under) an obligation is necessarily to be obliged. I am obliged to the person to whom I am under an obligation and obliged to do what I am under an obligation to do. I may, however, have an obligation or obligations—though I cannot be under an obligation—to someone according to which I am obliged to do something for him, though I am not obliged to him. Of this kind are treaty obligations, e.g. to come to the help of an ally in the event of an attack by a third power, or family obligations, e.g. to play with my children; or a doctor's obligation to his patients. I have obligations to my allies, my children and my patients, and am therefore, obliged to do something for them; though I am not obliged to them, nor under an obligation to them.

But though to have (or be under) an obligation is necessarily to be obliged, the converse is not true. I can be obliged without having (or being under) an obligation. Neither the yachtsman who is obliged by the wind to alter course, nor the traveller who is obliged by the gunman to hand over his wallet, nor the candidate who is obliged to take paper X, nor the chess player who is obliged to sacrifice his bishop, has (or is under) an obligation to do what he is obliged to do.

The essential difference, as H. L. A. Hart[2] has argued, between being obliged and having (or being under) an obligation is that the latter is restricted to those instances in which one is legally or morally obliged. Hence, he who has a legal or moral obligation is legally or morally obliged and *vice versa*, whereas there is no such thing as a physical or psychological obligation for one who is obliged by force of circumstances. The notion of obligation operates within the context of institutionalized rules. There will usually be a rule that anyone in such a legal or moral position as a promissor, debtor, contractor, parent, doctor or treaty power must do so-and-so in such-and-such circumstances; the position comprises, amongst other things, an obligation or set of obligations. These legal and moral obligations include social, family and treaty obligations. Such obligations may call for immediate action or for action only if and when certain circumstances arise; e.g. a nation's obligation to come to the help of its allies when attacked or a doctor's obligation to visit his patients now or when they become ill.

The fact that the notion of *obligation* is different from that of *being obliged* and that the former is confined to legal and moral contexts does not, however, show that the notion of being obliged in 'legally or morally obliged' is different from that in 'obliged by force of circumstances'. One must be careful, therefore, not to attribute to the fact that having an obligation differs from being obliged characteristics which are really due to the fact that having a (legal or moral) obligation or being legally or morally obliged, differs from being obliged by force of circumstances. Thus, Hart's supposition that the statement that a person had an obligation to report for military service may be true irrespective of his beliefs and fears of discovery and punishment, whereas, allegedly, the statement that a man was obliged to hand over his money to a gunman means that he did the latter because of his beliefs and fears does not prove a difference between 'obligation' and 'obliged' but only between 'obligation' and 'obliged by force of circumstances'. For a man may truly be said to be obliged (i.e. legally obliged) to report for military service independently of his beliefs and fears, since what he has a legal obligation to do he is legally obliged to do. Similarly, Hart's assertion that 'whereas the statement that he had this obligation is quite independent of the question whether or not he in fact reported for service, the statement that someone was obliged to do something normally carries the implication that he did it' depends for its truth on the type of factor by which one is obliged. One must have done what one was physically obliged to do, but not necessarily what one was legally or morally obliged to do. Prosecuting counsel might well say in 1965 of a doctor who was under contract from 1960 to 1962 that he was obliged by his contract to visit the deceased when telephoned, but that he failed to do so.

Equally mistaken is the common view that a difference in the implications of 'legally or morally obliged' and of 'obliged by force of circumstances' shows a difference in the meaning of 'obliged'.

For instance, one of Hart's arguments is essentially this: 'He was morally (or legally) obliged to V' implies 'He was under an obligation to V', whereas 'He was obliged by the gunman to V' does not imply 'He was under an obligation to V'; therefore, there are two senses of 'oblige'.[3] But this argument is no better than the similarly fallacious argument that because a man who is compelled by orders to retreat retreats 'under compulsion', whereas a man who is compelled by the heat of the flames to retreat does not retreat 'under compulsion', therefore there are two senses of 'compel', or that because 'The Simonstown Agreement requires us to V' implies 'There is a requirement under the Simonstown Agreement that we V', whereas 'Circumstances require us to V' does not imply 'There is a requirement that we V', therefore there are two senses of 'require'. A soldier who is ordered to shoot any looters on sight is under orders (has orders) to shoot looters on sight. A soldier on the parade ground who is ordered to pick up his rifle is not under orders (nor has orders) to pick up his rifle. But 'is ordered' is being used in the same sense throughout. Nor is the argument any better than the differently fallacious argument that because 'The butler was suspected' implies 'The butler was under suspicion', whereas 'Foul play was suspected' does not imply 'Foul play was under suspicion', therefore there are two senses of 'suspect'. The difference in implication between 'He was morally (or legally) obliged' and 'He was obliged by circumstances' can be explained by the kind of factor by which he was obliged. And similarly for 'compel' and 'require'. The difference in implication between 'The butler was suspected' and 'Foul play was suspected' can be explained by the logical difference between that on which the suspicion fell, e.g. the butler, and that of whose occurrence there was a suspicion, e.g. foul play. The syntactic differences in the sentences here can be brought out, e.g., by 'The butler was suspected of foul play' and 'Foul play by the butler was suspected'.

A second argument commonly used[4] to suggest that there are two senses of 'oblige' uses the premisses 'He was obliged by the gunman to V' implies 'He did V', but 'He was obliged by his contract to V' does not imply 'He did V'. The objection to this argument is that by parity of reasoning one would have to invoke two senses of, e.g., 'have to', 'require', 'demand'. For it could be argued, first, that 'Owing to the storm everyone *had* to abandon ship' implies that the ship was abandoned, but 'According to the regulations everyone *had* to have two sponsors' does not imply that everyone did have them; secondly, 'Shortage of oxygen *required* us to halt' implies that we did halt, but 'Our contract *required* us to deliver the goods within a fortnight' does not imply that the goods were delivered; thirdly, 'Circumstances *demanded* that he V' implies that he Ved, but 'Honour *demanded* that he V' does not imply that he Ved. Such an extension of ambiguity is fantastic. The solution lies not in two senses of 'oblige', 'have', 'require' or 'demand', but in the different ways in which different factors make things necessary. This is why an out-of-context use of the words 'He was obliged to do it' can leave one in doubt whether what they say implies that he did it and whether it implies that he was under an obligation to do it.

Incidentally, the absence of any implication between being legally or morally obliged to do something and doing it does not, as is sometimes alleged, show a difference between obligation and necessity. For the commonly adopted axiom of modal logic, 'It is necessary that p implies p', i.e. $(Lp \rightarrow p)$, has exactly the same exceptions. If force of circumstances makes it necessary that p, then it will be the case that p; but what is legally or morally necessary is no more bound to be than what is legally or morally obligatory. Conversely, it is a similar mistake to say[5] that the axiom of modal logic that what is actual is possible, i.e. $(p > Mp)$, has no analogy in deontic logic on the ground that what is actual is not necessarily permitted (or allowed). For we can say, first, that whatever is actual is permitted or allowed by physical circumstances—as when the lever permits the rod to move or the hole allows excess water to drain off—secondly, that much of what actually happens is not legally or morally possible, and, thirdly, that whatever is actually legal or moral is legally or morally permitted.

The ubiquity of the move from a difference in implication to a difference in meaning can easily be shown[6] in philosophical theories about other concepts, such as *responsibility*, *see* and *believe*. But, although a word does sometimes have two senses and although this is sometimes the explanation for differences in implication, too easy a recourse to this ploy deflects one's attention from the real source of the difference and, hence, provides only a temporary satisfaction.

(c) 'OBLIGED BY'

Having argued that common to all the uses of 'oblige' is the idea of being obliged by something, and also that having (or being under) an obligation implies being obliged, I want now to examine this notion of being obliged by.

When I am obliged by something, I am obliged by it either to someone or to do something. In both cases there is something I am obliged to do, but only in the latter case is this deed specified. Both the kindness of the man who opens the door for me and the inconsiderateness of the man who locks it against me (or the faulty mechanism which makes it stick) oblige me to do something, though the latter obliges me to do something which can be specified as being any alternative other than the one barred to me by the locked door, while the former obliges me only to make some unspecified return for the help given.

Though the only thing that can make me obliged to someone is a deed of his, there are many factors which can oblige me to do something specific. I can be obliged by persons, e.g. a gunman; by circumstances, e.g. a locked door, ill-health or the inefficiency of a servant; by my own deed, e.g. the giving of a promise or the assertion of a thesis; by my position, e.g., as a secretary or a doctor; or by rules and regulations, e.g., of morals, the law or a club. Sometimes the obliging factor is not mentioned, but only what I am obliged to do; e.g. to give my full name and address, to offer either examination paper X or paper Y, if I offer paper D.

Ordinarily a person of a certain position, in certain circumstances, with certain aims and following certain rules of conduct, has some freedom of choice. There are various alternative ways in which he can behave compatibly with his position, circumstances, aims and rules. Within this sphere he can do whatever he wants. A yachtsman intent on making port may have several alternative routes; there may be no agreed date on which a manufacturer has to deliver his goods, a candidate's choice of papers may be fairly unrestricted, a chess player may have several possible moves. In different circumstances, however, a factor may be present or may intervene which limits this choice by leaving him with no alternative, compatible with his general position, but to do so-and-so. He is now obliged by this factor to do so-and-so. This factor makes it necessary for him to do it. The wind obliges the yachtsman to alter course; the terms of his contract oblige the manufacturer to deliver his goods within a fortnight; the degree regulations oblige the student who offers paper D to choose either paper X or paper Y; a particular move by his opponent may oblige a chess player to move his king. As the case of the student shows, what one is obliged to do may be to choose within a restricted field rather than to take without choice. Similarly, the man who is under an obligation because of a kindness received or a promise given or who has obligations to his children or his allies is restricted, in a specified or unspecified way, in regard to his future acts. I may not be free to dine with you because I have promised to visit someone else; my country may have no choice but to go to war, if its allies are attacked.

What I am obliged to do is not something I cannot help doing or something I am unable not to do. I may sometimes quite easily not do what I am obliged to do. It is something I am unable not to do without incurring certain consequences. There is always an 'otherwise' in the offing.[7] I have to do what I am obliged to do, otherwise these consequences will follow. The manufacturer who is legally obliged by his contract to deliver the goods within a fortnight can only not do so at the cost of breaking the law. The man who promises to do X can only not do X by infringing some moral rule. Where one is obliged by force of circumstances, the 'otherwise' in the offing is the loss of some specific objective. The yachtsman who is obliged to alter course can only not do so by throwing away his chance of reaching port; the traveller who is obliged to hand over his wallet to the gunman can only avoid doing so by giving up his life. It is always possible to act incompatibly with the legal, or moral situation or with one's objectives. But given that there is no alteration in these, one is in the stated circumstances obliged to follow a certain course. The question whether someone has a choice or not does not occur in a vacuum, but in a particular situation. The question is whether, in this situation, I am obliged to do X or whether, physically, legally or morally, I may choose X or an alternative. It is a different question whether I have a choice between doing X and either giving up my objective or behaving illegally or immorally. What I am physically obliged to do is what, given adherence to physical conditions and to my objective, I have physically no option but to do. What I have a moral or legal obligation to do is what, given my situation and the moral or legal conditions, I have no moral or legal option but to do. The obliging factor narrows down what I can physically, legally or morally do. There is, of course, no suggestion that what legally or morally one is obliged or has to do is what physically one is obliged or has to do or that a narrowing of one's legal or moral options is a narrowing of one's physical options. What I am legally or morally obliged to do can conflict either with something else I am also legally or morally obliged to do or with something I am physically obliged to do, though I cannot be physically obliged to do conflicting things, since what I am physically obliged to do, unlike what I am legally or morally obliged to do, I in fact do.

Obligation share with its genus necessity—and with truth—the property called by logicians 'distribution through logical conjunction'[8] Thus, if I am obliged to give prizes to both A and B, then I am obliged to give a prize to A and obliged to give a prize to B. The notions expressed by 'best' and 'right', like those expressed by 'false', 'impossible', 'dangerous' and 'strange', do not have this property. It does not follow that because the best or right thing (or an impossible or dangerous thing) is to give A and B a prize, therefore the best or right thing (an impossible or dangerous thing) is to give A a prize and the best or right thing (an impossible or dangerous thing) is to give B a prize. Hence, one cannot give a correct analysis of *obligation* in terms of what is best or right.[9]

To be obliged to do something is not to have something done to one. The yachtsman is obliged to alter course, he is not swept off it. The gunman obliges me to hand over my wallet, he does not snatch it from me. To be obliged to do something is quite unlike being physically compelled to do something, even when the obliging factor is a physical event such as a fall of snow on the railway-line by which I had hoped to travel. For a physical compulsion, unlike an obliging factor, turns a deed done by me into something done to me. When I am obliged, I still have a choice and am still called upon to act, but the choice is not among alternatives at the same level as the obligatory alternative. At that level, I must take the obligatory one. The difference between what I do because I am obliged to do it and what I do because I choose to do it is not the difference between a choice and a free choice,[10] but between a choice whether or not to act inconsistently with one's general position and a choice among alternatives compatible with one's general position. If I am obliged to alter course, I do not choose this course from alternatives, though neither is it impossible for me not to take it. I alter course because, given that I do not want to be driven out to sea, I have physically no choice but to. The fact that I could choose to let myself be driven out to sea is irrelevant, since the original problem was whether, given that I did not want to be driven out to sea, I had any choice of course. Similarly, the manufacturer is obliged to deliver the goods within a fortnight because, given that he is to fulfil his contract, he has legally no choice but to deliver by that date. Hence, there is no paradox in saying that a man who is obliged to do X has no choice, even though it is physically possible for him—and perhaps quite easy—not to do X.[11]

What I do thinking I am obliged to do it, I do; it is not something done to me. Yet I do not do it voluntarily. To say[12] that I was obliged voluntarily to do so-and-so is a contradiction. I do not do it voluntarily because I do not choose to do it in preference to another alternative; the other alternative was, I thought, closed to me by the obliging factor. What I voluntarily do is contrasted with what I do thinking I am obliged to do it and not, except by philosophers[13] and jurisprudents, with what is done to me. A trade union member may give a voluntary subscription to charity, but he is obliged to subscribe to his union fund; a man may give himself up voluntarily or because he thinks he is obliged to do so; a student may withdraw voluntarily from the university or be obliged to withdraw. Contrast answering because you were asked a direct question and volunteering the information. The contrast between a voluntary and an obligatory subscription, surrender or withdrawal is not the same as the contrast of a voluntary payment and a levy, a voluntary surrender and a capture, or a voluntary withdrawal and expulsion. For the former is a contrast between one thing I do and another that I do, one voluntarily and the other obligatorily, while the latter is a contrast between what I do and what is done to me. What I do thinking I am obliged to do it, I do not voluntarily do; since I think I have to do it. What is done to me I do not voluntarily do, since I do not do it at all.

It is these two contrasts which, it seems to me, Aristotle is struggling to distinguish in the opening sections of the third book of the *Nicomachean Ethics*; i.e. (i) the contrast between throwing one's goods overboard to save oneself in a storm and throwing them overboard voluntarily, and (ii) the contrast between being carried out to sea by the wind and putting out to sea. He is also unclear, I think, on the difference between doing something involuntarily (or non-voluntarily) and doing it unwillingly.

Further, if what I do is unintentional, unknowing or in some respects non-attentive, it is neither voluntary nor obliged, though it is still a deed of mine. If, by a voluntary remark of mine, I quite unintentionally hurt your feelings, I did not hurt them voluntarily nor was I obliged to hurt them. If, by voluntarily handing over a letter to you, I give you a secret document by mistake, I did not voluntarily or non-voluntarily pass on secret information. What I do absent-mindedly is neither voluntary nor obliged.

The contrast between an act and a happening is one way in which *being obliged* is different from *being bound*. I can only be obliged to do something, whereas I can be bound either to do or to feel something or to have something happen to me. I can be bound, but not obliged, to succeed or to fail, to cry or to feel tired. I can be obliged to surrender, but only bound to be caught; obliged to jump, but only bound to fall. Since inanimate objects do not perform actions, they can only be bound to do what they do. The shopkeeper may be obliged to raise his prices, but his prices can only be bound to go up. If I am obliged to do so-and-so, I have no alternative to doing it other than acting incompatibly with my moral, legal or desired position. If something is bound to happen, there is no alternative to its happening unless some other part of the situation is altered. Thus, the inhabitants are bound to be massacred unless the relief column arrives before nightfall; it is bound to rain unless the wind changes. Whereas 'obliged' carries an 'otherwise' rider, 'bound' has an 'unless' rider.

Many philosophers[14] have suggested that the alternative which I am obliged to take must be less desirable, less desired, or less advantageous than the other alternatives; that 'the logic of obligation requires a conflict between the obligation to do something and the inclination not to do it'.[15] This is a mistake. Being in the position of being able to choose among alternatives, of being able to do whatever I want, is more desirable and perhaps more desired than being restricted in my choice by having to do something whether I want to or not. But it does not follow from this that the alternative to which I am restricted is less desirable or desired than the alternatives once open to me; nor that I cannot be said to be obliged to do what is advantageous to me. A candidate who wishes to offer paper D may be obliged to choose his further option from papers X and Y, but this need not be a less desirable course to him than choosing between papers P and Q. To be free to live either within or without the city boundary is more desirable than to be obliged to live within it; but there need be no suggestion that it is less desirable to live where one is obliged to live than to live elsewhere. The fact that exchanging a pawn for a queen or other piece is an advantage in chess does not preclude[16] this change's being obligatory in certain circumstances. It is simply false to suppose that 'to say to someone that he has an obligation to refrain from torturing children plainly implies that he would want to torture them if he had a chance'.[17] Often, of course, the obligatory action is less pleasing than the alternatives and is one which I would not have taken if I could have avoided it; but not necessarily so. Furthermore, even when the obligatory course is undesirable, it is not this quality which makes it obligatory, but the fact that it is the only course open to me in the circumstances. It is not the fact that I consider a visit to the dentist undesirable *per se* that allows me to say I am obliged to go,[18] but the fact that it is the only course open to a person who wants to stop his toothache or, perhaps, who has promised to go. The obligatory course is the course I have to take whether I want to or not; not the course I would not take unless I had to. It is not a question of whether I want the course or not; since the course is obligatory, that question does not arise. The fact that we often excuse ourselves by pleading that what we did was not what we wanted to do but only what we were obliged to do does not show[19] that being obliged to do something implies not wanting to do it. Nor is it relevant[20] whether the end to which the obliged alternative may be a means is pleasant or unpleasant. If I have only five minutes to get through my dinner before an appointment, I will be obliged to rush through the meal irrespective of whether the appointment is for an evening at the theatre or for a boring committee meeting. Nor is it necessary[21] for me to have an interest in or concern about keeping the appointment. It is keeping the appointment that contributes to the obligatoriness of my haste.

Nor should we confuse, as some philosophers have done,[22] the circumstances which may oblige me to do something and a general objective whose preservation may depend on my doing the obliged action. In virtue of my general objective I may have several alternatives open; I am not obliged to take any particular one of them. What obliges me to take one of them is that factor which closes the others. It is the wind, and not his desire to avoid being swept out to sea, which obliges the yachtsman to alter course. Certainly, I would not do the obligatory deed unless I wished to attain my objective, and sometimes it is necessary to mention what this objective is. Thus, I may be obliged to support a colleague on one measure on Senate because I want his support on another. But to say that I am obliged to do X if, or because, I want Y, or in order to get Y, is not to say that it is the desire for Y which obliges me to do X, or that I am obliged to do X by my desire for Y. My desire for Y is not the obliging factor. What does oblige me to do X is that factor which rules out any alternative ways of getting the desired Y. What obliges me to support my colleague on Senate is not the desire to have his support in return, but the fact or the knowledge that his support of me is dependent on my support of him. This fact or knowledge limits my choice of means of getting his support. I know that only by supporting him can I get him to support me. What obliges me to do X partly explains why I do X, but it is a mistake to suppose[23] that it is my reason or motive for doing X; or that being obliged to do X means doing X in order to get or preserve Y. I already have my general reason or motive for doing X, namely a desire for Y, which is a reason for doing whatever is a means to Y. What the obliging factor does is to restrict my choice of means. Motives or reasons do not oblige me to do anything. Hence, there is no need to try to distinguish obliging from non-obliging motives (e.g. Nowell-Smith) or the kinds of reasons for action that motives provide from the kinds of reasons that obligations provide (e.g. Gauthier).

Because the law actually attaches sanctions and penalties to failure to do what one is legally obliged to do, jurisprudents and moral philosophers have sometimes[24] identified the obliging factors in law with the sanctions, and supposed that to say that one is legally obliged to do so-and-so means that it has to be done in order to avoid unpleasant consequences. But this is a mistake. Sanctions are not legal bonds; they are measures employed to strengthen these bonds. To be physically obliged, e.g. by a fall of snow, to take one rail route rather than any other is to have the alternatives physically closed, thus making it a physical impossibility to take any alternative other than the obligatory one, though not, of course, physically impossible to abstain from action altogether. To be legally obliged, on the other hand, e.g. by railway regulations, to take one route rather than any other is to have the alternatives only legally closed. There need be no physical difficulty in taking an alternative other than the legally obligatory one. Hence, the law usually introduces sanctions and penalties to prevent people from doing, or to punish them for doing, what it is legally, though not physically, obligatory not to do. But reference to such a sanction is no part of the meaning of 'legally obligatory'. Similarly, any reference to the part played by social pressures or qualms of conscience in getting people to do what is morally obligatory is quite irrelevant to the meaning of 'morally obligatory'.

Among moral philosophers the notion of being obliged is often assimilated or confused with the notion expressed by ought. I shall try to show in my discussion of *ought* in the next chapter that both in morals and elsewhere *ought* and *obliged* are quite distinct.

NOTES

1. Harré wrongly analyses *necessity* as a kind of obligation.
2. E.g. (1961), chapter v, § 2; Baier (1966), p. 220a, note 24, asserts that 'being obliged cannot be used to explain having an obligation'; but his assertion lacks argument.
3. Kearney, p. 47 suggests that Hart meant only to distinguish 'oblige' and 'obligation' and not two senses of 'oblige'. Page (1972) and (1973) certainly uses this argument for two senses of 'oblige'.
4. E.g. Page (1972) and (1973).
5. E.g. von Wright (1968), p. 93.
6. Cp. White (1971).
7. Cp. Nowell-Smith (1954), p. 202.
8. Cp. Castañeda (1965), pp. 337–44; von Wright (1951b).
9. Cp. Castañeda (1968), pp. 141–2; Bales, pp. 203–5.
10. E.g. Nowell-Smith (1954), p. 202; cp. Gauthier, p. 176.
11. Cp. Nowell-Smith (1954), p. 201; Hart and Honoré, p. 72.
12. E.g. Gauthier, p. 176.
13. E.g. Aristotle, *N.E.* III, 1; Nowell-Smith (1954), p. 201; Gauthier, p. 176; Ryle, pp. 73–4.
14. E.g. Nowell-Smith (1954), pp. 206 ff.; Hart (1961), pp. 80, 85; Gauthier, pp. 177–9; Baier (1966), p. 212a.
15. Nowell-Smith (1954), pp. 210–11; cp. Kant.
16. *Pace* Gauthier, p. 177.
17. Baier (1966), p. 212a.
18. E.g. Gauthier, p. 178.
19. *Pace* Nowell-Smith (1954), p. 206.
20. E.g. Gauthier, p. 178.
21. E.g. Page (1972) and (1973).
22. E.g. Gauthier, pp. 179 ff.; cp. J. Austin, *Lectures on Jurisprudence*, xxii–xxiii.
23. E.g. Gauthier, p. 183; Nowell-Smith (1954), p. 204; Hart, p. 80.
24. E.g. J. Austin, *Lectures on Jurisprudence*, xxii–xxiii; Nowell-Smith (1954), pp. 200, 204, 209, 242.

Chapter Ten

Ought

(A) OUGHT

'Ought'—and its variant 'should'—can, like most auxiliaries, be followed by an almost unlimited range of verbs[1]* irrespective of whether such a verb indicates an act (e.g. move, kill) activity (e.g. be playing, be climbing), passivity (e.g. be killed, be given) or acquisition (e.g. discover, gain), a possession (e.g. have, know), state (e.g. believe, enjoy, be asleep), feeling (e.g. be ashamed, be glad), attitude (e.g. want, despise) or ability (e.g. be able, be capable) or any kind of occurrence (e.g. fall, happen).

As with 'must', there are, however, two importantly different ways in which 'ought' modifies its main verb. 'Ought to V' may indicate either subjunctively 'ought to be that it (should)[2] V' or indicatively 'ought to be that it does V'. Hence, the syntactical ambiguity of the sentence 'There ought to be a law against it' as said, on the one hand, by the recent victim of a confidence trick and, on the other, by a student looking through the Statute Book, or of 'The teachers ought to get a rise in salary', as, on the one hand, a prescription and, on the other, a prediction. Because almost any example of 'ought to V' could, with sufficient ingenuity, be taken either way, only the context will show which is meant. Normally, 'If one is tired, one ought to take a rest' would be an example of the subjunctive-governing use and 'If one takes a rest, one ought to wake refreshed' of the indicative-governing use.

The fact that the subjunctive use of 'ought to V' differs from the indicative use has various consequences. The indicative use works for any verb, but the subjunctive usually works only for

* References to this chapter begin on p. 162.

task-verbs, e.g. the indicative 'He ought to discover it' as contrasted with the subjunctive 'He ought to look for it'. Advice and preference are linked to the subjunctive use; discovery and prediction to the indicative. The reasons appropriate to the subjunctive are preferential, those to the indicative are evidential. This is why 'He knew that legally and morally he ought to have declared the contraband' makes sense, whereas 'He knew that legally and morally he ought to have been able to conceal the contraband' is nonsense.

It does not follow, however, that these two uses of 'ought to V' are due, as the OED and some philosophers[3] suggest, to two senses of 'ought', e.g. 'ought' as 'right' or 'better' and 'ought' as 'highly likely' or 'naturally expected', any more than that the difference between 'It is possible that this should be so' (or 'It is possible for this to be so') and 'It is possible that this is so' is due to two senses of 'possible' or that the difference between an expression of intention and a prediction by the use of the words 'I will come' is due to two senses of 'will'. The current half-truth that 'the word "ought" is used for prescribing' is no more indicative of the meaning of 'ought' than the opposite half-truth that the word 'ought' is used for predicting. In both the subjunctive-governing use and the indicative-governing use 'ought' has exactly the same sense, which can be explained as follows. Any problem about what ought to be—like any problem about what can, may, need or must be or what is best, right, obligatory, or whatever—arises in a situation identified in terms of a set of circumstances, a requirement, a set of alternatives and an aspect. What ought to be is, as its etymology in several languages suggests, what amongst the alternatives is *owing* in these circumstances and under this aspect in order that the requirement be met. It is as if the situation were a pattern with one missing piece, namely, what ought to be.

In the indicative use of 'ought to V' the only kind of requirement is conformity to the facts and the only aspect is the factual. Since however, the relation expressed by 'ought', unlike that expressed by 'must', is not one of necessity, but of what is owing, what ought to be is that which follows non-deductively from the given or presupposed circumstances. For example, if the train left London at 10 o'clock at its usual speed, at what time ought it to arrive in Hull? If he usually works late at the office, then he ought to be there now. If the square root of 900 is 30, then the square root of 837 ought to be about 29.

In the subjunctive use of 'ought to V', on the other hand, the kinds of requirement, whether loosening a nut, getting to London or keeping healthy, and the kinds of aspect, whether technical, academic, legal or moral, are legion. Because of the variety of requirements from which and the aspects under which the circumstances may be thought to have something owing in the subjunctive use of 'ought to V', as contrasted with the one kind in the indicative use, 'ought' in the former, but not in the latter, is often qualified both by such phrases as 'in order to', 'if you want to . . .' and by such adverbs as 'technically', 'morally', etc. One cannot, however, assume that the absence of such qualifiers is an infallible sign of the indicative use, since they are often omitted in the other use when the context is supposed to make the type of requirement or aspect clear. It is, indeed, because this supposition is often false that misunderstandings arise as to what are the grounds on which and the aspects from which it is being asserted that something ought to be Ved.

Given the circumstances and the requirement, reasons can always be demanded for the contention that what is owing is so-and-so. Such reasons will consist in a claim that certain features either of the circumstances or of the requirement or of one of the alternatives make this alternative what is in the circumstances owing in order to meet the requirement. Each kind of reason is a complement to the other kinds. For example, you ought to take a rain coat because the clouds look threatening and/or because you do not want to spoil your new suit and/or because a coat will keep you dry. You ought to be ashamed of yourself because of something dishonourable you have done and/or because honour demands it and/or because such a feeling would do you honour. This nut ought to do because crank-case bolts are of a standard pattern and/or because it came from another crank-case.

Although a reason why is always relevant to the contention that something ought to be, and, hence, the notion expressed by 'ought' implies the relevance of a type of reason, it would be a mistake[4] to suppose that this reason is any part of the meaning of 'ought' or that the notion that something ought to be is to be analysed in terms of *a reason for it.*

(B) OUGHT, RIGHT AND BEST

Attempts have been made to analyse *ought* in terms of some supposedly more basic concept. Two frequent favourites are an analysis in terms of *right*[5] and an analysis in terms of *best.*[6]

That *best* and *right* are different concepts is clear enough. Common to both is the idea of selection from a set of alternatives on a certain basis; the difference being in the kind of relation of the selected alternative to the basis of selection. The best X is the X which comes nearer than any other X to possessing or possesses in a higher degree than any other X a set of characteristics taken as a standard for X's. The other X's may be good or bad X's and some of them may be better than others. The right X, on the other hand, is the X which meets a particular specification. The other X's are wrong X's however near they come to meeting the specification. Even a bad X may be the best available X, but a wrong X cannot be the right X. Something may be *less* than the best, but *other* than the right. A specification may, of course, be such that the best, or only the best, meets it and is, consequently, the right X. Conversely, the right X may be the best X. But neither of these connections is necessary, except in the weak sense that the right X may be the best X for meeting the specification, and *vice versa.* If someone is looking for my house, then no. 33 is the right house, but no question arises in the circumstances whether it is the best house. On the other hand, a particular house can be the best house I ever saw or the best available or possible, but it makes no sense to call it the right house I ever saw or the right house available or possible.

Neither the idea of possessing the desirable characteristics to the highest degree (the best) nor the idea of meeting the specification (the right) is the same as the idea of being what is owing in the situation (ought); though both being what is best and being what is right can be reasons for being what ought to be. Depending on the situation, what ought to be may either be what is best or what is right or what is both. 'How should one address an archbishop?' asks for the right mode, whereas 'How should one address a lecture class?' asks for the best mode. Someone ought to be ashamed of what he has done because shame is the right or proper emotion to feel in the circumstances; someone ought to change his ways because changing his ways is the best thing to do. The reason why something is said not to be as big, wide or dark as it ought to be may be either that there is a right size, width or shade or that it would be better to be bigger, wider or darker.

It is the difference of what ought to be from either what is best or what is right that explains why 'It ought to be that p and q' implies and is implied by 'It ought to be that p and it ought to be that q'—since if *p and q* is owing, then *p* is owing and *q* is owing—whereas 'It is best (or right) that p and q' does not imply nor is implied by 'It is best (or right) that p and it is best (or right) that q'—since p and q cannot both be the best or the right thing.

I don't think that the idea expressed by 'ought'—any more than the idea expressed by any other modal—can be analysed into or explained by any more basic ideas; but a rough and ready equivalent might be the idea of *being appropriate*.

(c) OUGHT AND IS

Saying that something *is* clearly is different from saying that it *ought* to be. Neither implies nor in any way involves the other. Anyone who supposed that whatever ought to be is would be merely a prey to wishful thinking, while anyone who supposed that whatever is ought to be, besides being an ultra-conservative, would lay himself open to Hume's just charge of overlooking the fact that the relation expressed by 'ought' is entirely different from that expressed by 'is' and 'cannot be a deduction' from it.[7]

Equally mistaken is the view[8] that 'A ought to have Ved', whether used in the indicative or the subjunctive way, either implies, presupposes or suggests that A did not V. Like the view that 'A could have (or might have) Ved' implies, presupposes or suggests that A did not V, it rests on a confusion of the *propriety* of saying something in certain circumstances with the *truth* of saying it. Neither 'He ought to have done it by now' nor an affirmative answer to the question 'Do you think he ought to have done it?' either implies, presupposes or suggests that he did not do it.

What has, however, interested and puzzled philosophers is how the assertion that some *one* thing ought to be is related to the assertion that some *other* thing is, e.g. how the assertion that someone ought to stop smoking is related to the assertion that smoking is a cause of lung cancer or—though this type of example has not usually been discussed—how the assertion that someone ought by now to have reached his destination is related to the assertion that the distance is quite short.

We have already seen that the assertion that one thing *is* is commonly and popularly given as a reason for the assertion that another *ought* to be; that is, that one thing ought to be because another is. Furthermore, I have argued that the features which constitute reasons why one thing ought to be can be features either of the circumstances or of the requirement or of the chosen alternative. Such features make what ought to be what is owing in the situation. It is clear, therefore, that to argue that one thing ought to be, e.g. that my friend ought to give up smoking or ought to arrive soon, because another thing is, e.g. smoking causes cancer or the journey is short, is to furnish an elliptical argument which would have no force without its implicit assumption of a certain set of circumstances and a certain requirement viewed under a certain aspect. A misapprehension about these is a common cause of disagreement about what ought to be if and when such-and-such is. Hygiene, but not economics, may make the liability to cause lung cancer a reason why one ought to give up smoking; but in a peculiar situation where lung cancer destroyed some more malignant growth even hygiene would not make smoking inappropriate. Serious practical problems arise when one's worry about what ought to be done is due to a prior worry about which requirement, e.g. health or economy, ought to be taken into account, for this prior worry can only be cured by adopting a higher requirement from which one considers which lower order requirement is appropriate. In addition, what morally ought to be need not be the same as what legally ought to be.

The logical problem is how the assertion that a particular set of circumstances or a particular requirement has such-and-such a feature is related to the assertion that so-and-so ought to be. The assertion that one thing ought to be cannot be logically implied by the assertion that something else is because it is not necessarily true to say that, if one thing is, another thing ought to be. It is not even necessarily true that if smoking causes lung cancer, one ought from the medical point of view to give it up. Likewise, it is not necessarily true on logical grounds that if someone set off half an hour ago on what is normally a ten-minute journey, he ought to have arrived by now.

To say that none of these hypotheticals is necessarily true is not, of course, to say that none is certainly true nor that none can be proved to be true, for a contingent hypothetical can be as certain and as easily proved as a necessary hypothetical. 'If oil is added to water, the two will not mix' is certainly true and has been proved to be true, but it is not necessarily true. And if a hypothetical, whether contingent or necessary, is or is proved to be true, then if the antecedent is or is proved to be true, the conclusion is or is proved to be true.

There can be no doubt that a host of hypotheticals of the form 'If X is so, then Y ought to be so'—and a corresponding collection of consequences of the form 'Y ought to be so'—though not necessarily true, are certainly true and can be proved to be true: 'If one is tired, one ought to take a rest' and 'If one takes a rest, one ought to wake refreshed'; 'If a wire has to bear a load of more than ten tons, it ought to be three inches thick' and 'If it is three inches thick, it ought to bear a load of more than ten tons'; 'So-and-so is bigger, heavier, darker, more expensive, than it ought to be'. Experience, together with some general principles, proves, though it does not imply or mean,[9] that certain kinds of actions and of happenings are, from the logical, hygienic, economic, strategic or technological point of view, or for the sake of safety, completeness or tidiness or some specific interest, what is owing or appropriate in certain kinds of circumstances.

Another common way of supporting the assertion that something ought to be is by an appeal to accepted standards of conduct. Thus, we say that something ought to be done as a matter of courtesy, tact or honour, in fairness, in all conscience or honesty. The suggestion here is that the rules of courtesy, fairness, etc., act like the principles of hygiene, economics or common-sense in making so-and-so what is owing or appropriate in such-and-such circumstances. Here again, though it is not necessarily true that, e.g., in fairness or in courtesy such-and-such is what one ought to do—for this follows neither from the meaning of 'fairness' (or 'courtesy') nor from the meaning of the words used to describe the features of the situation[10]—it may be quite certain that this is what ought to be. It may also be certain that only actions within a certain range, partly indicated by the meanings of the words, could even be candidates for the position here. There are, in the current jargon,[11] internal relations between certain requirements and certain ideas of what ought to be. It would need a very special explanation to show, e.g., how one could defend the remark that in all fairness, or as a matter of courtesy, one ought to tax a citizen according to the number of forenames his father had.

To argue that because a set of circumstances, a requirement or an alternative has such-and-such features, therefore this alternative is the one which ought to be in these circumstances is like arguing that because an engine driver watches out for the signals, bends, crossings and other traffic on his line or because a surgeon watches out for the breathing, heart condition and blood pressure of his patient, therefore each is a careful worker. There are certain characteristics of the engine driver's job and certain characteristics of the surgeon's job which, it is generally agreed, make these or those features in it worthy of attention and, therefore, make the paying of attention to them the hallmark of a careful worker. But it is not part of the meaning of 'careful', 'engine driving' and 'surgical operation' that just these features are worthy of attention. The concept expressed by 'ought' is one of a huge number of concepts which are related to their instances neither purely accidentally nor strictly necessarily, but in relation to the acceptance of certain rules and principles by which the criteria of their application are decided. Other such concepts are those expressed by 'valid', 'real', 'eligible', 'careful' and, of course, 'right' and 'good'.[12]

(D) OUGHT AND CAN

Several features of our everyday thinking suggest that we can't always do what we ought.

For instance, we frequently don't do what, admittedly, we ought to do. And like anything else we don't do, sometimes we don't do it although we can do it and sometimes we don't do it because we can't do it. We often don't do what we could have done because we are forgetful, lazy or unwilling. When, however, we don't do something because we could not do it, what prevents us from doing it may be some physical circumstance, like lack of facilities or time, or some psychological circumstance, like the temporary weakness of will which afflicts us all or the permanent disability from which, e.g., a kleptomaniac suffers. Because of lack of money the tired worker cannot take the holiday he ought to take; nor can the patient help touching the sore which he knows he ought to leave alone. Similarly, failure to do one's duty, to carry out one's obligations, to live up to one's ideals, to fulfil one's promises, or to do what is best or right can be due simply to not doing any of these or to being unable to do them. Many people have to strive to do their duty and some find that they strive in vain. 'I couldn't do it' is often offered as an excuse for not doing what we ought to do. But an excuse would be unnecessary if it weren't what we ought to do.

Sometimes one not only does not, but cannot, do what is right or what is best just as sometimes one not only does not but cannot do what one wants or would like to or even what one needs to or what is necessary. A man or a company may suddenly find that it cannot pay its debts. There are such things as unattainable standards and impossible conditions. Frequently, in order to attain a desired end, certain means or methods ought to be used which, for some reason or other, cannot be. If, as I have argued, what ought to be is what is 'owing' in the circumstances, it is not surprising that what ought to be need not always either be or be possible. It is not unusual or odd that someone cannot do what he ought to do, particularly what he thinks he ought to do. What would be odd, as Socrates and Hare have emphasized—indeed over-emphasized—is for someone simply not to do what he thought he ought to do if he knew he could quite easily do it.

The characteristics in virtue of which something is what ought to be done are different from the characteristics in virtue of which it is something which can be done. There is no reason why the two sets of characteristics should coincide. The reasons of health because of which we ought to give up smoking or leave a fresh wound alone are different from the psychological reasons why we find these impossible. It is also because what we ought to do is not necessarily something that we can do that we have occasion to use such an expression as 'We ought to deal with this matter now, if we can'. The qualification 'if we can', like the qualification 'if we have time', would be simply redundant if we ought to do only what we can.

Our feelings are notoriously something we cannot control. Yet circumstances are often such that we ought to feel surprised, perturbed, ashamed, grateful, anxious, more helpful, etc., when in fact we not merely don't feel any of these, but can't help feeling the opposite. It is not uncommon to suggest that there is something positively wrong, that is, not as it should be, with someone who does not and cannot feel in certain ways in certain circumstances. And however ridiculous, unreasonable or even neurotic it may be to torture oneself with feelings of guilt or remorse for not doing what one thought one ought but couldn't, it is not an impossible situation to be in. Indeed, Aristotle (*Nicomachean Ethics*, 1150b–29) distinguished the man who deliberately didn't do what he ought from the man who couldn't help not doing what he ought as the man who didn't from the man who did feel remorse.

Nor is it only what ought to be done or what ought to be felt that often can't. What ought to be or ought to happen also often can't.[13] Indeed, when what ought to happen doesn't, it is usually —or, if determinism is right, necessarily—because it can't. This is why when so-and-so which ought to have happened doesn't we say that something must have gone wrong. The gas which fails to flow where it ought to flow is usually being prevented from doing so. When someone who ought to be here by now isn't here, we wonder what's keeping him.

In addition to all these features of our thinking which suggest that one can't always do what one ought, there are several logical absurdities which result from its denial.

For instance, if one gets oneself into the position where one ought to do something and also ought to do something inconsistent with it, then it seems to follow that one ought to do both. But it certainly does not follow that if it is possible to do something and also possible to do the opposite, then it's possible to do both. Hence, what ought to be need not be what can be.

Secondly, if what we ought to do could only be what we can do, we would have too easy a way of escaping our commitments, neglecting our duties and welshing on our obligations. We would simply do something that made it impossible for us to do what we ought to do and, therefore, on this view, made that no longer something we ought to do. For instance, most of us think that we ought to return books we have borrowed. Does anyone think that by destroying borrowed books and, thus, making it impossible to return them, we have made it no longer true that we ought to return the borrowed books?

Thirdly, it is not implausible to hold—as perhaps do Socrates and R. M. Hare—that a man who sincerely thinks he ought to do so-and-so and can do it will do it. If, therefore, someone who thinks he ought to do so-and-so does not do it, it follows that he can't do it. The only way to avoid this conclusion and, therefore, to uphold the traditional dictum that one can always do what one ought to do would be to deny the original principle.[14] But the original principle seems at least as plausible as the traditional dictum.[15]

Fourthly, several philosophers, including G. E. Moore,[16] have suggested that 'One ought to feel so-and-so' is really an abbreviation for 'One ought to feel so-and-so if one can'. But this suggestion is open to the same illegitimate infinite regress as was Moore's analogous suggestion that 'One can' is an abbreviation for 'One can, if one chooses'. For the 'ought' of 'ought to feel so-and-so, if one can' would itself—since it is an 'ought' which qualifies a feeling verb—have to be expanded to 'ought to feel if one can' and so *ad infinitum.* Further, one could not employ the usual *modus ponens* to argue 'I ought to feel so-and-so, if I can; I can; therefore I ought to', since this last 'ought to' could only mean 'ought to, if I can'.

In contrast to the deep-rooted common-sense belief—confirmed also by Aristotle's discussion[17] of *akrasia* (incontinence)—that one cannot always do what one ought stands a distinguished line of moral philosophers from Kant to Hare,[18] and including Sidgwick, Moore, Broad and Ross, who have held that *ought* implies *can*, that to have an obligation to do something implies having the power to do it and that nothing can be a man's duty unless he is able to do it. We have here a conflict reminiscent of the conflict between the common-sense view that often one could have done something other than one did and the philosophical belief in determinism, or the conflict between the philosophical view that what has happened cannot be improbable and the common-sense belief that the improbable sometimes happens. Can this conflict be resolved?

One common attempted reconciliation which will not work is to suppose, as do Sidgwick[19] and Moore, that there are two senses of 'ought',[20] one of which implies and the other of which does not imply *can*. This solution is advocated because it is admitted[21] that there are certain things which one ought in certain circumstances to feel even though one's feelings are something which one cannot control. Moore, for instance, takes the last of the Ten Commandments, 'Thou shalt not covet thy neighbour's house, nor his wife, nor his servant, nor his ox nor his ass, nor anything that is his', as an assertion that one ought not to have certain feelings. And, as we saw, it is often true that we ought not to be surprised, over-anxious or too hopeful about so-and-so or that we ought to feel ashamed, grateful or pleased about such-and-such.

This solution, however, will not work since, although it allows that there is a sense of 'ought' in which one cannot always feel what one ought, it denies that there is any sense of 'ought' in which one cannot always do what one ought; and, therefore, it does not reconcile the philosophical tradition and the common-sense belief. Furthermore, it offers no reason, other than its alleged ability to explain failing to feel what one ought, for supposing that there is a second sense of 'ought'. Nor, as I have argued, can any good reason be given for this supposed second sense.[22] Similar objections apply to an attempt by Broad[23] to distinguish 'You ought to V' as implying factual possibility from 'It ought to V' as implying logical possibility; for it too denies that one can't do what one ought.

A second unsuccessful attempt at reconciliation introduces what it calls two *uses* of 'ought'. R. M. Hare,[24] for instance, distinguishes an alleged *prescriptive* use of 'ought' in which one can always do what one ought from a *descriptive* use in which one cannot. The descriptive use, moreover, is denigrated as being 'off colour' or 'down-graded'. One objection to this solution is that a prescriptive use of 'ought' is partly defined as a use in which one can always do what one ought and is, therefore, not evidence in favour of such an assertion. A second, more important, objection is that Hare has confused the use of a word, e.g. 'ought', in virtue of which it has or lacks certain implications with the use to which someone might put a statement expressed by a sentence containing that word, e.g. 'ought'. What Hare states, quite correctly, is that it would be pointless to tell someone that he ought to do something which it was admitted he could not do, if the purpose of telling him that was to give him some (moral) advice or guidance. Hence, one cannot speak 'prescriptively' and also ask the impossible. But this does not show that in any but some 'off colour' sense of 'ought' it must be that someone always can do what he is told he ought to do, much less that he could have done what he is told he ought to have done. Indeed, in his discussion of moral weakness, Hare himself shows some awareness of the difference between the view that being prescriptive necessitates not asking the impossible and the view that 'ought' in any but a 'down-graded' sense implies *can*. He is more anxious to show that the existence of moral weakness is not an exception to the former view than that it is not an exception to the latter. Hence, he says, on the one hand, 'the typical case of moral weakness . . . is a case of "ought, but can't" ' and on the other, that 'typical cases of moral weakness do not constitute a counter-example to prescriptivism'.[25]

A third type of solution, advocated for instance by Hare, is that the relation of *ought* to *can* is not that of implication, but of some weaker sort, such as the relation of presupposition which, according to Frege and Strawson, the statement that the King of France is wise has to the statement that there is a King of France. But this solution avoids none of the clashes between the traditional philosophical view and the commen-sense belief. For I don't presuppose, any more than I imply, that I can keep my tongue off the sore tooth or that I can avoid feeling elation when I tell myself I ought to.

Although both Moore and Hare do try to some extent, by the introduction of two senses or two uses of 'ought', to reconcile the common-sense belief that not everything that ought to be can be with the philosophical tradition that *ought* implies *can*, the main tenor of their views is to deny the former, if what it asserts, as it undoubtedly does, is that in a plain straightforward sense one cannot always do what one ought. Other contemporary philosophers certainly deny this.

One of their reasons[26] is that it is pointless and unreasonable to insist that someone ought to do what he cannot. But, as we have already seen, the supposition that this is pointless and unreasonable does not show that it is either incorrect or meaningless. Secondly, there is the often stressed fact[27] that one does not blame or punish someone for not doing what he ought to do if it is discovered that he could not do it, either because of some physical circumstances such as lack of facilities or some psychological impediment such as mental weakness. But the fact that one does not blame or punish him does not show that one does not believe that what he failed to do was something he ought to have done. We don't blame him because we accept his inability as an excuse. But it wouldn't be an excuse for not doing what he ought to do if it could make that cease to be what he ought to do. Thirdly, in a practical situation the discovery that one cannot do what one ought to do may lead one to ask what should one do now or instead. But this, *ex hypothesi*, does not show[28] that the original was not something one ought to have done. Thus, as chairman of tomorrow's meeting, I ought to read thoroughly through all the papers before-hand. If, for whatever reason, I leave myself so little time that I cannot do so, then I ought at least to glance at them. But the fact that as a chairman pressed for time I ought at least to glance at them does not show that as a chairman it is not the case that I ought to read them thoroughly. The question what I ought to do as a chairman is quite different from the question what ought I to do as a chairman pressed for time and, therefore, the answer to the first question may allow something which the answer to the second question does not, e.g. read the papers thoroughly.

This last distinction enables me to lead into my own solution to the apparent incompatibility of the common-sense belief that one can't always do what one ought with the philosophical tradition that *ought* implies *can*. This solution is to distinguish between what is internally and what is externally impossible, that is, between what is impossible in the case stated and what is impossible because of something else. What is correct in the philosophical tradition is that what ought to be in certain circumstances is what, as far as these circumstances are concerned, can be. Thus, on the one hand, the chairman of tomorrow's meeting ought to read thoroughly beforehand all the agenda papers because it is perfectly possible for such a thing to be done. Since, on the other hand, it is not logically possible for a chairman with insufficient time, nor empirically possible[29] for a chairman with, say, only half an hour to spare, to read thoroughly beforehand all the papers, it cannot be true that such a thing ought in these new circumstances to be done. Similarly, it can be true that a man ought to return borrowed books because that a man return borrowed books is a logical and empirical possibility, but it cannot be true that a man ought to return destroyed books, since this is not possible. In other words, those circumstances which make so-and-so something which ought to be done must also allow it to be something which can be done. Nothing can be made by the same circumstances both what ought to be done and what cannot be done.

What, on the other hand, is correct in the common-sense belief that one can't always do what one ought is that there may be additional circumstances, either physical or psychological—such as lack of time or lack of will power—which make what ought to be done, and what apart from these circumstances can be done, into something which cannot, because of these additional circumstances, be done. I can't because of lack of time, look through the papers which, as chairman, I ought to look through, nor can I, because they have been burnt, return the books which, being borrowed, I ought to return. In other words, when one can't do what one ought to do, additional circumstances make impossible what is in itself something that both can be and ought to be, just as, when one does not do what one ought to do, additional circumstances prevent, though they do not make impossible, the occurrence of what is in itself something that can be and ought to be. If I have several jobs which ought to be done this afternoon, then the fact that for some reason I cannot do them does not turn them into jobs which no longer ought to be done this afternoon, though it is necessarily true that jobs which cannot be done this afternoon are not jobs which ought to be done this afternoon. Philosophers who deny the common-sense belief about 'ought' and 'can' have slipped from the everyday truth that in some circumstances one can't do what one ought to do to the tautology that it is not the case that one ought to do what, because of the circumstances, one can't do. They have imported the description of the circumstances which make the deed impossible into the description of the deed to be done. But the false supposition that there can be an X which is both what ought to be and cannot be differs from the fact that there ought to be an X which there cannot be. A parallel distinction is this. What makes it probable or improbable that something is so need not make it either the case that it is so or the case that it is not so. Hence, there is no incompatibility of its being improbable that something is so with its being so. The chances against a man's winning a lottery are not altered by the factors that produce his win.

A failure to observe this distinction between what is impossible in itself and what is impossible for some extra reason accounts for Hare's denial or denigration of the common-sense belief that one can't always do what one ought. This denial is based on the correct principle that 'if a description of an action is such as to rule out a practical "Shall I?" question, then it will also rule out, for the same reason, the corresponding universally prescriptive "Ought I?" question'. Hare's examples show that this means that where it makes no sense to ask whether something can be done, it makes no sense to say that it ought to be done. Thus, we cannot ask 'Shall I?'—meaning 'Shall I do?'—about falling downstairs by accident, being driven ashore, going to the wrong room by mistake.[30] Since this is not something it is possible to do, it is not something one ought to do. That this is the correct interpretation is borne out by the fact that it makes good sense to ask whether these are things that ought to happen—in the way that it makes sense to ask whether the gas ought to come out through this hole —because these are things which can happen. It is not, as Hare suggests, because it is prescriptive or used to answer a practical question that *ought* implies *can.* Nor, conversely, would a proof that 'ought' is not always used prescriptively or to guide action show that it did not imply 'can'.[31]

The fact that what it makes no sense to do cannot be something one ought to do does not, however, show that so-and-so can only be something that ought to be done if, in circumstances additional to those that make it something that ought to be done, it also is something that can be done. For instance, it does not follow that this can be a job that I ought to do only if, despite my tiredness, it is a job I can do. Hare has slipped from the plausibility of denying that 'ought to V' is applicable where 'V' expresses some kind of impossibility to the fallacy of denying that 'ought to V' is applicable where for some extra reason it is impossible to V. Another way of bringing this out is as follows. Hare also says that 'It is, in fact, the impossibility of deliberating or wondering whether to do a thing which rules out asking whether one ought to do it.' It is clear that he means that anything about which it does not make sense to deliberate or wonder cannot be something about which it makes sense to ask whether it ought to be done. But this is not because it is something which for some reason or other cannot be done, it is because it is not a deed at all. This is why he also describes it as something which 'could not be the subject of a decision, order, request or piece of advice'.

But saying that so-and-so can only be something that ought to be done if it is something which can be described as 'something done' differs importantly from saying that so-and-so can only be something that ought to be done if it is something which, in the particular condition of the agent or in the circumstances of the case, can be done. The former view is part of the necessary truth encapsulated in *ought* implies *can*—I say 'part' because the dictum covers not merely 'ought to and can do' but also 'ought to and can feel, be, happen', etc.[32]—while the latter is what the common-sense belief that one can't always do what one ought to do wishes to deny.

This distinction between what is impossible in itself—or impossible under a given description—and what is impossible because of additional circumstances explains why philosophers so plausibly maintain that, when faced with something that is, and is supposed, known or discovered to be, impossible, one cannot then consistently and solemnly pronounce that it ought to be done. Common-sense, on the other hand, stresses that one often can't do what one ought to do because normally the question what one ought to do or whether one ought to do so-and-so comes up and is answered before we learn or realize that we can't, for whatever reason, do it. This is perhaps why[33] there is less of an air of paradox about saying 'I know I ought to do it, but I just can't' than saying 'I clearly can't do it, but I ought to'. For in the former case the course of action is described without reference to its impossibility, whereas in the latter the knowledge of the impossibility accompanies the description of the deed and is, therefore, implicit in it. The sad but legitimate complaint that a man cannot always do what he ought should not be interpreted as entailing the logical absurdity that a man ought to do what he cannot do. To adapt Lewis Carroll, 'You might just as well say "I can't what I ought" is the same as "I ought what I can't."' An analogy is the plausibility of saying that something is 'improbable but true' and the apparent paradox of saying 'It is true but most improbable'. We only considered its probability because we didn't know its truth or falsity just as we only considered whether it ought to be done because we didn't know its impossibility.

Our main distinction may also explain the logical paradox that in a conflict of duties, obligations, etc., we may find that we ought to do so-and-so and ought also to do something inconsistent with it, whereas we cannot both do so-and-so and also do something inconsistent with it. For such a conflict is not usually an internal inconsistency, whether logical or empirical, in the very description of the double task, such as that between paying one's taxes and not paying one's taxes or between getting a job done in time and taking a lot of time off, but an external inconsistency caused by some feature of the situation, such as the frequent inconsistency between discharging one's duties as a busy doctor and discharging one's duties as an attentive father. The philosophical dictum that *ought* implies *can* rightly disallows the internal inconsistency while the common-sense belief that one often cannot do what one ought allows the external inconsistency.

Incidentally, even if it were correct to argue that 'A ought to do X' implies 'A can do X', it would be a mistake[34] to suppose that the contrapositive of this—namely, that 'A cannot do X' implies 'It is not the case that A ought to do X'—is an example of the implication of *ought* by *is*. 'It is not the case that A ought to do X' is not an example in which it is said that anything ought or ought not to be done; to suppose otherwise is to confuse 'It is not the case that A ought to do X' either with 'A ought not to do X' or with 'A ought to do not-X'.

(E) OUGHT AND MUST

Our earlier examination of 'must'—and its variation 'has to'—showed that it has many of the characteristics which we now see are typical of 'ought'. First, like 'ought', it can accompany almost any verb to indicate either, subjunctively, that it must be that something be so or, indicatively, that it must be that something is so. Only the context will show whether, e.g., 'He must meet the champion' means that it is necessary that this should be (or for this to be) or that it is necessary that this will be. With 'must', but not with 'ought', this difference is sometimes shown by the use of a variant verb—in this case 'has to'—as in the contrast between, e.g., 'The wound had to be cauterized' and 'The wound must have been cauterized'.

Secondly, like 'ought', 'must' indicates a relation of what must be to a situation consisting of a set of circumstances and a requirement viewed under a certain aspect. For the indicative-governing use the requirement is conformity with the facts and the aspect logical; for the subjunctive-governing use each may be of many kinds. Hence, 'He had to do it', unlike 'He must have done it', can be qualified by 'legally', 'morally', 'by regulation', or whatever. Furthermore, this variability in the kind of requirement or aspect allows that what one had to do—e.g. legally or morally—one need not in fact have done; whereas what one must have done, one did.

Thirdly, the existence of the relation expressed by 'must', like that expressed by 'ought', is supportable by reasons whose function is to show that either the circumstances or the requirement or what must be has features which link the last to the others. For example, it might be suggested that the victim must have fallen from a great height because of the type of fractures he sustained and that the suspect had to be locked up because he was violent, but had to be released later because of *habeas corpus.*

In addition to these similarities, the relation expressed by 'must' is like that expressed by 'ought' in that it is quite different from and logically undeducible from that expressed by 'is'. Thus, although it may be quite certain that, according to the law, if a car is put on the road, it must be insured for third party risks, this is no more a necessary truth than is the equally certain truth that, if the brain is damaged, the mental functions must be impaired.

It may be partly because of such similarities that philosophers, and particularly moral philosophers, have assimilated 'must' and 'ought'. This they have done in one or other of three ways: *either* quite explicitly[35] *or* implicitly, that is, by using the two concepts indifferently and interchangeably,[36] *or*, as we shall see in the next sections, by correctly identifying what one must do with what one is obliged to do and incorrectly identifying what one is obliged to do with what one ought to do.[37]

The notion expressed by 'must' is, however, quite different from that expressed by 'ought'. Whereas to say that something ought to (or should) be is to say that it *owes* it to or is appropriate to the situation to be; to say that something must (or has to) be is to say that it is *necessary* in the situation for it to be. What ought to be is in the situation the appropriate, perhaps because the right or the best, thing; what must be is the only thing. For instance, if wet emery paper is more effective than dry as a rust remover, then it ought to be used for that purpose; but if it is the only effective means, it must be used. To say that what ought to be must be is too strong, while to say that what must be ought to be is too weak. Only a difference either in the circumstances or the requirement can turn what ought to be into what must be or what must be into what ought to be. Furthermore, whereas the subjunctive-governing use of 'must' (in many tenses expressed by forms of 'have to') is like that of 'ought' in that it does not imply actuality, the indicative-governing use of 'must', unlike that of 'ought', does do so.

(F) OUGHT AND OBLIGED

Being obliged is a species of being necessary that there should be (or necessary for there to be). Hence, what someone was obliged to do is what he had to do, not what he must have done and, therefore, not necessarily what he did. Hence, also, we can be obliged in manifold ways, whether morally, legally, economically, by regulation, etc. As its etymology suggests, 'obliged' signifies that what is obliged is tied to the circumstances and requirement because of which it is obliged.

These connections between 'obliged' and 'must' show that the very common philosophical identification[38] of 'obliged to do' (and 'obligation') with 'ought to do' is the same as the identification of 'must' with 'ought'. It is a mistaken identification. 'Obliged', like 'must', indicates, as we saw, that because of a particular requirement, whether legal, economic, prudential, or moral, the features of the situation are such that the agent has only one course open to him; the obligatory course is necessary because it is the only course. 'Ought', by contrast, indicates that in the situation one of several courses open is the right or the best course.[39] Given certain circumstances and a requirement, no question of what he ought to do arises for the yachtsman who is physically obliged by the wind to change course or for the manufacturer who is legally obliged by his contract to deliver the goods by a specified date. Advice can be given to someone in the form of a statement about what he ought to do because such a statement differentiates between the merits of alternatives; but to tell someone what in the circumstances he is obliged to do is only to give him information about which course is the only one open to him.

Where one has conflicting ends which would oblige one to take contradictory courses, one can ask oneself from a different point of view which end, if any, one either ought to or is obliged to pursue. Thus, I may know that if I try to climb a hill, its steepness will oblige me to change gear and also know that if I am to save my gearbox, its faulty condition will oblige me to stay in the same gear. I have, therefore, to ask myself whether I ought or am obliged to try to climb the hill and thus damage the gearbox or to try to save the gearbox and thus fail to climb the hill. Where one has conflicting legal or moral obligations, one is legally or morally obliged to do both of the things one has an obligation to do. Hence, from a different point of view the only question can be which of the two one ought to do. Thus, my promise to my wife may oblige me to be home for dinner, while my duty as a doctor may oblige me to stay late at the hospital. Ought I to leave the hospital early to keep my promise and so fail in my duty as a doctor or stay at the hospital to do my duty as a doctor and so break my promise to my wife? Only from some other point of view can I find out what I ought to do.

Any attempt[40] to discover a necessary connection between a statement about what one ought to do and a statement about what one is obliged (or under an obligation) to do is doomed to failure. It involves *either* the sensible, but admittedly too weak, supposition that because according to one requirement, e.g. of a promise made or a desire for economy, one is obliged to do so-and-so, therefore, when all things are taken into consideration one ought to do it; *or* the nonsensical, because too strong, supposition that because according to one requirement, e.g. of a promise made or a desire for economy, one is obliged to do so-and-so, therefore, by the same requirement, one ought to do it. Although the same thing can be for one reason something I ought to do and for another something I am obliged to do, as when I am obliged because of my office to do what I ought as a matter of courtesy to do anyway, *ought* and *obliged* are, in relation to one and the same reason, mutually exclusive—though not exhaustive—since the former implies that there are, while the latter implies that there are not, alternative courses of action. This is why we can set at rest anyone's worry whether he has done the right or the best thing by assuring him that he has done the only thing

(G) MORALLY OUGHT

The logical characteristics of *ought* (and of *must* and *oblige*) do not change when the concept is used in a moral setting. Indeed, there is no reason to suppose that the sense of any verb is changed by a change in its qualifying adverb. The sense of 'ought' (or of 'obliged') in 'morally ought (or obliged)' no more differs from its sense in 'economically, aesthetically, legally, academically, ought (or obliged)', then the sense of 'ought' in any of these differs from its sense in, e.g., 'temporarily, unofficially, allegedly, certainly, ought (or obliged)'. The common philosophical talk[41] of a moral sense of 'ought' (or of 'obliged') is either a simple mistake or a misleadng way of distinguishing the sense of the whole phrase 'morally ought' (or 'morally obliged') from that of, e.g., 'economically ought' (or 'economically obliged').

What does change is the aspect under which and the kinds of reasons for which things are regarded as being what ought to be (or what must be or are obliged to be) done. What makes something either the right or the best or the necessary or the 'owing' moral course may be quite different from what makes it either the right or the best or the necessary or the 'owing' practical, economic, logical or legal course. The great uncertainty about what we ought to do (or are obliged to do) in moral predicaments stems largely from an uncertainty or disagreement about what constitutes the principles of a moral point of view. Fortunately, the resolution of this uncertainty is not our business here. Further, since moral judgements are limited to what it is right, best, necessary, or whatever, that there should be (or for there to be), there is no place in morals for that use of 'ought' (or 'must') which indicates that so-and-so is what it ought to be (or must be) that there is. We cannot sensibly say that morally speaking the wound ought to have (or must have) healed by now.

The notion that morally speaking one *ought* to do so-and-so differs from the notion that morally speaking one *must* or *is obliged* to do it in the same way as that in which these notions differ in non-moral contexts. Depending on the features of the situation and on our moral code, there will be some things one ought to do and others one is obliged to do.[42] Status, position, office, duties and commitments normally carry with them certain obligations, while the features of alternative courses and actions normally show one of them to be the appropriate one, the one we ought to choose. To hold that one ought, rather than that one is obliged, to keep one's promises sounds too weak; whereas to hold that one is obliged, rather than that one ought, to go to the help of the traveller set upon by thieves sounds too strong. At the very least, it suggests a different conception of the moral life. It is because some moral philosophers[43] have subscribed to the very general principle that morally the only thing one can do is that which, all other things being considered, is the right or the best thing to do that they have made what one is obliged to do and what one ought to do necessarily coincident. But to say that the morally owing (or the right or the best) course—what we ought to do—is the only (or necessary) moral course—what we are obliged to do—is not to utter a tautology. The idea of being owing—morally or otherwise—that is, the idea expressed by 'ought', and the idea of being necessary—morally or otherwise—that is, the idea expressed by 'obliged', are distinct ideas.

NOTES

1. The only exceptions I can find are either other auxiliaries or some other verbs which take an infinitive, e.g., deserve, mean, intend to V. The 'should' of 'ought' should not be confused with the 'should' of 'shall'.
2. This is the 'should' of 'shall'.
3. E.g. Montefiore, pp. 28, 31; Frankena, pp. 157–75; Gauthier, pp. 10–12. Contrast Wertheimer, chs. 3, 4 and 5.
4. E.g. Edgley, § 4.10; D. G. Brown, § 3.3.
5. E.g. Samuel Clarke and Richard Price; Broad (1930), pp. 164–5; Ross (1930), ch. 1 and (1939), ch. III; cp. Nowell-Smith (1954), ch. 13; Ewing (1947), pp. 123–4; Hare (1952), ch. 10. G. E. Moore seems to have distinguished 'ought' from 'right' by means of the superlative and the comparative of 'good'—see the passages mentioned in White (1958), pp. 144–7.
6. E.g. Montefiore, p. 39; Baier (1958), ch. 3; Diggs, p. 302; Gauthier, pp. 10–12; G. R. Grice, ch. 1; Sloman, pp. 385–94; Margolis (1970), pp. 44–53; (1971), chs. 3 and 4.
7. *Treatise* III.i.1; Although Hume is here discussing moral philosophy, I do not think that his remarks about the logic of *is* and *ought* need to be confined to morals.
8. Lyons, p. 309.
9. Contrast Black (1964), pp. 165–81.
10. Cp. Hare (1964).
11. Cp. Foot (1959).
12. Cp. White (1964), pp. 5–6 and the references there to Ryle and Wittgenstein; Anscombe (1958); Wertheimer, pp. 150 ff.

13. *Pace* Robinson (1971a), pp. 193–202, whose denial of this is due to the mistaken view that 'ought' here means 'It is possible'.
14. E.g. Kordig, pp. 11–12.
15. So Gilbert, pp. 141–4 tries to save the general principle by weakening the traditional dictum to mean that ought implies being capable in general rather than able in the circumstances. On this distinction, see Henderson, pp. 108–9.
16. (1922), pp. 315–23; cp. Frankena.
17. *Nicomachean Ethics*, VII, 1–10.
18. Cp. Prichard (1932), p. 6; Frankena; Hampshire (1951), pp. 161–78; Smith, pp. 362–75. The suggestion of Frankena and of O'Connor (1971), p. 26 that 'I believe I ought' implies 'I believe I can' equally contradicts the common-sense belief that, e.g., it's often true that 'I know I ought to do it, but I don't think I have time'.
19. *Methods of Ethics* (1907), Bk. I, ch. III; cp. Broad (1930), p. 141; (1952), pp. 195–217; Frankena, pp. 160–1.
20. Smith alleges several different senses of both 'ought' and 'can'; cp. Robinson (1971a), who holds that 'prudential' and 'probable' *ought* imply, while 'moral' and 'ideal' *ought* do not imply, *can*.
21. Contrast Ross (1939), pp. 52–5.
22. Even less so for four or more senses, as Frankena and Smith suggest.
23. (1930), pp. 141–2.
24. (1963), chs. 4–5; cp. Hampshire's distinction between 'moral judgments' and 'appraisals', 'expressions of ideals', etc.
25. This sensible separation of the two points is, unfortunately, inconsistent with Hare's general position which partly defines the prescriptive use of 'ought' in terms of '*ought*' implies '*can*'.
26. E.g. Hare, Hampshire and O'Connor; contrast W. G. Maclagan in the symposium with Hampshire and Hare.
27. E.g. Montefiore, p. 25; Smith, p. 367.
28. *Pace* Ross (1939), p. 109; Smith, pp. 367–8.
29. My position is quite different from another of Broad's, which distinguishes between logical and empirical possibility (1930), pp. 141–2.
30. Or 'blinking', e.g. Montefiore, p. 25.
31. *Pace* Margolis (1971), pp. 40, 42, 60.
32. *Pace* Margolis (1971).
33. Contrast the explanation of Wertheimer, pp. 129–30 that it is an instance of the principle that the greater excludes the less.
34. E.g. Mavrodes; contrast Shaw, pp. 196–7.
35. E.g. von Wright (1963a), ch. 8; (1963b), ch. 5, § 12; Baier (1958), p. 283; Black (1964), p. 170.
36. E.g. Moore (1912), pp. 34 fl.; Prichard (1949), *passim*; Nowell-Smith (1954), *passim*.
37. E.g. Prichard (1949), *passim*; Nowell-Smith (1954), *passim*; Anscombe (1958), *passim*; Hare (1963), p. 170; Smith; Grice.
38. E.g. Prichard (1949), pp. 89–91; Hare (1963), p. 170; von Wright (1963); Baier (1958); Black (1964); Grice; Zink, ch. 4. Ewing (1947)

equates 'obligation' with what he calls one sense of 'ought'. Contrast Nowell-Smith (1954), ch. 14 and Gauthier, ch. 12. Ross (1939), chs. 3–4 distinguishes 'right' and 'obligatory', but seems to equate 'obligatory' both with 'the right thing' and with 'ought', cp. (1930), p. 3. Sesonske's criticism of the obligation/evaluation assimilation expressly denies any correlation of this difference with specific linguistic differences.

39. Many philosophers slip from 'best' to 'only' or 'necessary' in their analyses of *ought*; e.g. von Wright, Black, Gauthier, Diggs.
40. E.g. Searle, pp. 180–1.
41. E.g. Prichard (1949), pp. 90–1, *et passim*; Frankena; contrast Gauthier, pp. 18–23.
42. Cp. Hart (1958) and Sesonske.
43. E.g. Moore (1903), § 89, 'the assertion "I am morally bound to perform" is identical with the assertion "This action will produce the greatest amount of good in the universe" '.

Chapter Eleven

The Nature of Modality

In the preceding chapters I have investigated the particular behaviour of individual modal notions and hinted at a few of the ways in which they underline traditional philosophical problems. The general nature of modality should also have become clear, but it is appropriate to draw together in the final chapter some of the threads of the argument.

The basic question is what and how do the modals qualify. There have traditionally been two, not necessarily incompatible, views—often called '*de dicto*' and '*de re*'—about the nature of the subject of modal qualification. Though there is, as we shall see, some dispute about how exactly ancient and medieval logicians understood these views to differ from one another, the latin names suggest that on one view the modals qualify what is said about something and that on the other they qualify the something which it is said about. This is clearly how the difference is taken in some recent philosophy. For instance, von Wright's formulation[1]* is: 'modalities are said to be *de dicto* when they are about the mode or way in which a proposition is or is not true' and '*de re* when they are about the way in which an individual thing has or has not a certain property'.

I want to suggest that this distinction is based on two assumptions. The first, and correct, assumption is that modal qualifications can apply either to the whole or to part of something. The second, and incorrect, assumption is that what is qualified when the whole is qualified is a proposition, that is, something which is said. The difference between these two assumptions is roughly, though not exactly, equivalent to that between two Aristotelian and medieval contrasts; namely, the contrast between

* References to this chapter begin on p. 179.

an assertion taken in a *sensus compositus* and an assertion taken in a *sensus divisus* and the contrast between an assertion *de dicto* and an assertion *de re*.[2]

As regards the first assumption, one can distinguish further a genuine contrast of whole and part from an apparent contrast of whole and part. An apparent contrast of whole and part exists between, e.g., 'It is possible (necessary, certain, etc.) that X is Y' and 'X is possibly (necessarily, certainly, etc.) Y'. That this is only an apparent contrast between a qualification of the whole *X is Y* and the part *Y* is clear from our earlier detailed consideration of the modals when we saw that the adverbial form here is merely a variation on the impersonal 'It is F that . . .' form. 'X is possibly Y' is only another way[3] of saying 'It is possible that X is Y' just as 'He was surprisingly, unfortunately, conceivably or clearly very rude' is equivalent to 'It is surprising, unfortunate, conceivable or clear that he was very rude'. Medieval logicians sometimes called the impersonal form '*cum dicto*' and the adverbial form '*sine dicto*'.

A genuine contrast of whole and part, on the other hand, exists between, e.g., 'He possibly (necessarily, certainly) gave a misleading answer' and 'He gave a possibly (necessarily, certainly) misleading answer'. This contrast also occurs with many non-modal qualifications, as in the difference between 'He surprisingly, unfortunately, conceivably or clearly, gave a misleading answer' and 'He gave a surprisingly, unfortunately, conceivably or clearly misleading answer'. The existence of the form 'He gave a possibly (necessarily, certainly) misleading answer' does not, however, provide any evidence for a peculiar composite property *being possibly* (*necessarily*, *certainly*) *misleading* in addition to the property *being misleading*, any more than the existence of the form 'He gave a surprisingly (unfortunately, conceivably or clearly) misleading answer' provides evidence for a peculiar composite property *being surprisingly* (*unfortunately*, *conceivably or clearly*) *misleading*. Nor does the *de re* interpretation of modality imply the existence of such a peculiar property. The *de re* interpretation is not that 'X is necessarily (possibly) F' implies the existence of a property *being necessarily* (*possibly*) *F*.[4] It is that that which 'necessarily (possibly)' qualifies is X's being F and not the truth of the proposition—much less the proposition itself—that X is F, either in the ordinary sense or in the technically philosophical sense in which a proposition can be necessarily (possibly) true. The supposition of such a peculiar modalized property forces von Wright[5] into the *ad hoc* introduction of a special principle of predication to avoid certain paradoxes inherent in it.

More important, however, is the realization that the adverbial forms 'possibly (necessarily, certainly)' in 'He gave a possibly (necessarily, certainly) misleading answer' are as much variations on 'It is possible (necessary, certain) that . . .' as they are in 'He possibly (necessarily, certainly) gave a misleading answer'. The difference is that the impersonal 'It is possible (necessary, certain) that . . .' in the former qualifies only part of what is recorded, namely, that the answer was misleading; while in the latter it qualifies the whole, namely, that he gave a misleading answer. In exactly the same way, 'He gave a surprisingly (unfortunately, conceivably, clearly) misleading answer' qualifies the given answer as being one of which it is surprising (unfortunate, conceivable, clear) that is was misleading; while 'He surprisingly (unfortunately, conceivably, clearly) gave a misleading answer' qualifies the misleading answer as being something which it is surprising (unfortunate, conceivable, clear) that he gave.

Since the adverbial form of the modal is equivalent to the impersonal 'It is F that . . .' form in both the genuine and the apparent contrast of whole and part, we cannot properly distinguish the adverbial form as *de re* and the 'It is F that . . .' form as *de dicto*. This would be a rather inept way of making even the purely grammatical point that adverbs are predicated of verbs and adjectives, while expressions of the form 'It is F that' are predicated of whole clauses. We could, however, use[6] the contrast of whole and part—which some medievals may have called a contrast of *sensus compositus* and *sensus divisus*—to explain a difficulty of scope for which some medievals[7] may have wrongly used the contrast of modality *de dicto* and modality *de re*. This is the difficulty raised by the fact that 'It is possible for a man to write while he is not writing' is false, while 'It is possible for a man who is not writing to write' is true. The former is false because it allows the conjoint existence of two incompatibles, while the latter is true because it only allows the existence of one incompatible and the possibility of the existence of the other. Symbolically, the first is the illegitimate $M(p.\bar{p})$, while the second is the legitimate $\bar{p}.Mp$.

The equivalence of the adverbial form 'possibly (necessarily, certainly)' and the 'It is possible (necessary, certain) that . . .' form shows that the basic question about what the modals qualify centres on the second assumption that to say 'It is possible (necessary, certain) that X is Y' is to qualify the proposition, that is, what is said in, X is Y; an assumption which would make all modality *de dicto*.

The view that what is qualified when it is said that 'It is possible (necessary, certain, etc.) that X is Y'—or, as it is commonly put, 'It is possible that p'—is a proposition stretches at least as far back as Diodorus,[8] was common in medieval logic, and remains almost universal among contemporary philosophers.[9] In our earlier detailed examination of the concepts of *possibility*, *probability* and *certainty* we drew attention to its presence in the fashionable assumption that what is probable or certain when it is probable or certain that p is the proposition that p. An extreme version of it occurs in N. Rescher's[10] extended view of modality according to which not only the so-called alethic modalities (It is possible that p), but also his epistemic modalities (e.g. It is known or, expected that p), his temporal modalities (e.g. It was yesterday the case that p), his boulomaic modalities (e.g. It is hoped or feared that p), his deontic modalities (e.g. It ought to be brought about that p), his evaluative modalities (It is a good thing that p), his causal modalities (The existing state of affairs will bring it about that p) and his likelihood modality (It is likely that p) all exemplify the qualification of a proposition. Rescher[11] and many others follow Russell[12] in extending this thesis to the notion of *belief*, so that what is believed when it is believed that p is alleged to be the proposition that p. Indeed, Russell himself at various times used it, under the umbrella of 'propositional attitude', to cover desiring, assuming, judging, expecting, remembering, hoping, fearing, etc., that p. The main basis for such an assumption, is an unnoticed slide from the correct position that what is possible (necessary, certain), like what is hoped or feared and like what is surprising or unfortunate, is often *expressed or stated* by a proposition to the incorrect position that what is possible, feared or surprising *is* a proposition. Formal logicians are probably also tempted to assume that modal operators qualify propositions since they naturally look in modal logic for parallels to the propositional calculus, where we undoubtedly do have operators, true and false, whose arguments are propositions or propositional variables.

It is, no doubt, semi-conscious misgivings about their basic assumptions which often lead philosophers to move from the assertion that what is modally qualified, when it is said to be possible, probable, obligatory, or whatever, that p, is the proposition that p, to the view that what is qualified is the truth of the proposition that p. This leads to the substitution for 'It is possible (necessary, certain) that p' of 'It is possible (necessary, certain) that it is true that p' or, what is equivalent, 'It is possibly (necessarily, certainly) true that p'. Thus, Rescher substitutes[13] for 'p is permitted' the expression 'so acting as to render it true that p is the case is permitted' and quotes[14] approvingly an author who translates a causal statement into 'the conditions expressed are causally sufficient to make the statement describing the occurrence of the event true'. Even von Wright, who begins[15] correctly by stating that the deontic modalities (e.g. is permitted or obligatory) are about the way in which we are permitted or not to perform an act, soon moves[16] to the view that because 'possibly being A' denotes a property, but 'It is permitted to do A' expresses a proposition, therefore these deontic modalities are only to be taken *de dicto*, that is, as qualifying propositions.[17] But the relation between, e.g., 'He took an obligatory (permitted) course' and 'He took a course which it was obligatory (permitted) that he (should) take' is exactly the same as that between 'He took an unnecessary (possible) course' and 'He took a course which it was unnecessary (possible) that he (should) take'.

To say 'It is possible (necessary, certain) that the proposition is true' or, what is the same, 'The proposition is possibly (necessarily, certainly) true' is no more to say that it is the proposition itself which is possible (necessary, certain) than to say 'It is possible (necessary, certain) that the gearbox is faulty' or, what is the same, 'The gearbox is possibly (necessarily, certainly) faulty' is to say that it is the gearbox itself which is possible (necessary, certain). Likewise, to say that it is feared or hoped that the proposition is true is no more to say that the proposition itself is feared or hoped than to say that it is feared or hoped that the gearbox is faulty is to say that the gearbox itself is feared or hoped. 'Possible proposition' and 'necessary proposition' are, as we saw, logicians' jargon for propositions which are necessarily or possibly true, not in the sense that propositions can be possibly or necessarily known, unprovable, mistaken, existing or misleading—that is because of certain circumstances—but in the peculiar sense that they are in themselves possibly or necessarily true. It is a fundamental mistake to deduce from this technical use of 'possibly or necessarily true' certain conclusions about the general nature of modality.

An additional reason for the usual philosophical emphasis on the possibility or necessity of the proposition that X is Y being true, rather than on the possibility or necessity of X's being Y, may have been the false assumption that because 'X is possibly (necessarily) Y' is logically equivalent to 'It is possibly (necessarily) true that X is Y', therefore the former is the same as the latter. Not only is this assumption mistaken, but it can be lead to the mistaken and more dangerous conclusion that 'X is necessarily Y' is the same as, or is logically equivalent to, 'It is necessarily true that X is Y' in the technical philosophical sense that the proposition that X is Y is a necessary truth. This conclusion is dangerous because it involves us in such obviously fallacious paradoxes as the view that, if the number of the planets is necessarily less than eleven—which it is because the number of the planets is nine and nine is less than eleven—therefore, the proposition that the number of the planets is less than eleven is a necessary truth; which it is not.

I contend, therefore, that what is qualified when it is asserted, e.g., that it is possible, necessary, probable, certain that X is Y or that X is possibly, necessarily, probably, certainly Y, or when it is asserted that it is obligatory or permitted that X be Y or that X is obligatorily or permissibly Y, is not the proposition that X is Y or the proposition that X be Y. Furthermore, what is qualified when it is said that it is possible that such-and-such a proposition is true, or that such-and-such a proposition is possibly true, is not a proposition, but a proposition's being true. And the latter, unlike the former, is no more a *dictum* than the proposition's being known, unprovable or misleading. In the sense discussed, there is no such thing as modality *de dicto*. As we saw in detail, a wide variety of things can be qualified by different modals; but it is a variety which can all be classed as *de re*. The danger of the thesis that modality is *de dicto* is that it tempts one, particularly with such modalities as necessity, possibility, probability and certainty, to embrace subjective theories of modality according to which modality is a characteristic of thought rather than of that which can be thought about. It was apparently[18] on these grounds that Frege dismissed modal distinctions from logic.

A recent theory which has, on the one side, an unconscious kinship with the *de dicto* view and, on the other, a claimed relationship to Hume's subjective analysis of *necessity* as 'existing in the mind, not in objects', holds that 'modal concepts in general express the pressure of reason for or against doing things which people do'.[19] It holds, for example, that 'the word "must" expresses that pressure in which there is conclusive reason for someone to do something (including the case in which "do" is a stand-in for "think" or "believe"); the word "ought" that pressure in which there is good (but not necessarily conclusive) reason to do something; and so on'.[20]

According to this theory, in such an expression as 'Smith must (or ought to) hang his hat there', when it is used in what in earlier chapters I called the indicative-governing use—namely as meaning that it must or ought to be the case that Smith will hang his hat there—the modals *must* and *ought* do not apply, as grammar would suggest they do, to the action or happening itself, that is, to Smith's hanging his hat there, but to our thinking that there will be such an action or happening. The modals here, it is alleged, really signify that there is conclusive or good reason for thinking that Smith will hang his hat there. Thus, while *de dicto* theories transferred modality from actions and happenings to propositions about them, this theory transfers it from actions and happenings to our thoughts about them.

The exemplar actually claimed, however, by the advocates of this theory is not any holder of the *de dicto* theory, but Hume. It is said[21] of him that it was 'his merit to have recognised that such a displacement [viz. the grammatical displacement of modal words from thoughts about happenings to happenings] occurs with the concept of causal necessity'; though his particular account of it is rejected undiscussed.

This proposed variation on Hume, however, amounts in fact to the *de dicto* position, as we have seen it in, e.g., Rescher's statement of causal modality. For it is suggested[22] that Hume's thesis that 'the impression from which the idea of necessary connection derives is an internal one, either of, or generated by, the transition of the mind to the idea of the effect' should be put as 'the "must" of causal necessity expresses the relation of the conclusion that the effect occurred, is occurring or will occur, to its empirical grounds when these grounds are empirically sufficient to exclude any alternative conclusion'. Here clearly *de dicto* talk about 'grounds' and 'conclusion' has been substituted for *de re* talk of what is referred to in the grounds and the conclusion.

What is said[23] to be a similar variation on this same theme gives as the reason for the alleged grammatical displacement of the modals from their claimed association with the verb 'think'—e.g. 'I must (ought to) think that Smith will hang his coat there'—to their actual association with the verb denoting that action or happening which the thinking is about—e.g. 'Smith must (ought to) hang his hat there'—what is called 'the transparency of belief'. What this amounts to is an assimilation of 'I think that X is Y' and 'X is Y' on the grounds of the existence of the well known pragmatic paradox involved in saying 'I think that X is Y, but X is not Y' or 'X is Y' but I do not think that X is Y'. Because of this paradox it is alleged[24] that 'my present thinking, in contrast to the thinking of others, is transparent in the sense that I cannot distinguish the question "Do I think that p?" from a question in which there is no essential reference to myself or my belief, namely, "Is it the case that p?" '. We have frequently seen in the details of the earlier chapters that it is precisely this confusion between *the inappropriateness of saying* certain things, such as 'I'm not certain that X is Y, though X is Y', 'X is probably Y, though X is not Y', 'X may be Y, though it is not' and 'Ought I to do what I think is wrong?' and the *possible truth* of such things which is a main source of subjectivist views about probability, certainty, morality, etc. Another reason for the false supposition that the modals should be transferred from the action or happening to our thought about the action or happening may be a failure to see that in, e.g., 'Smith must (ought to) hang his hat there' the subjunctive-governing use asserts that it must (or ought to) be that something (should) occur, while the indicative-governing use asserts that it must (or ought to) be the case that something occurs. The view being discussed, therefore, rightly sees that, in the indicative-governing use, it is not an action or happening which is being modally qualified; but wrongly supposes that it is our thought about it. What is in fact being qualified is its being the case that there is such an action or happening.

Hume's own reason for asserting that 'necessity is something that exists in the mind, not in objects' is given[25] in terms of his theory that every idea must arise from a corresponding impression. He says 'there is no impression conveyed by our senses which can give rise to that idea [sc. necessity]. It must, therefore, be derived from some internal impression, or impression of reflection. There is no internal impression which has any relation to the present business, but that propensity, which custom produces, to pass from an object to the idea of its usual attendant. This, therefore, is the essence of necessity.' A parallel is drawn by Hume with logical necessity. 'Thus, as the necessity which makes two times two equal to four or three angles of a triangle equal to two right ones, lies only in the act of the understanding, by which we consider and compare these ideas; in like manner the necessity or power, which unites causes and effects, lies in the determination of the mind to pass from the one to the other.'

Hume's objection is essentially the same as the more recent argument that since there is no ostensive quality or object called 'necessity' in the way that there is an ostensive quality called 'red' or an ostensive object called 'table', 'necessity' must be the name of something subjective. Like its modern versions, it is applied equally to modal notions and to moral notions. It is not surprising that Hume held that 'morality is nothing in the abstract nature of things, but is entirely relative to the sentiment or moral taste of each particular being'. Moral ideas, like modal ideas, are founded on internal impressions.[26]

There is no need to repeat here the well-known objections to views which suggest that either 'good' or 'necessary' is the name of some subjective feeling or mental experience which I have in certain circumstances; that, e.g., 'good' signifies a feeling of approval or that 'necessity' signifies a feeling of compulsion and 'probability' a feeling of confidence. It is more important to note that subjectivism shares with objectivism a common mistaken assumption. This assumption, common to the Humean treatment and its modern variations, of both modal and moral notions, is that there are only two alternatives; either that such notions signify a physical quality or that they signify a mental quality—what Hume distinguishes as an external or an internal impression. Arguing correctly that they do not signify a physical quality in the way that, e.g., 'red', 'round' or 'old' does, it is supposed that they must signify a mental quality. But it is the assumption that such motions signify ostensive isolatable qualities, whether mental or physical, which is mistaken.

A recognition of the fallacy in such an assumption is the source of the view held by some modern grammarians and philosophers that what modal concepts do is not to *name* anything, either physical or mental, but to *express* something, such as the attitude or mood of the user of the concepts or to *indicate* an illocutionary potential. A typical grammarian's analysis is that modal words 'mark sentences according to the speaker's commitment to the factual status of what he is saying (his emphatic certainty, his uncertainty, doubt, etc.)'.[27] Similarly, a recent philosophical description[28] of what actual functions our modal terms perform calls them 'qualifiers of our assertions', a view which is exemplified in the case of 'It is probable that p' by saying that it is used to make a guarded assertion that p. Coupled with this illocutionary view of modal terms, there is a clear allegiance to a *de dicto* interpretation of their application. Not unexpectedly, proponents of this theory about modals hold similar theories, though without the *de dicto* overtones, about value concepts.[29] Indeed, illocutionary theories about value concepts, whether in a crude emotive form or in the more sophisticated guise of prescriptivism, antedate the corresponding theories about modals. Even in Hume's case, it has been argued[30] that his analysis of moral concepts is the source of his general philosophical views.

The illocutionary theory has the merit of seeing that there are more ways in which concepts can apply than simply to either objective or subjective properties and objects. But its demerits are, first, that it does not allow enough for the objectivity implied in our use of modal and evaluative concepts, that is, for such facts as that, though people may legitimately express contradictory views, only one of these views can be correct, that we do normally characterize assertions in which modal and evaluative concepts are used as being true or false, and that we do normally apply modal and evaluative concepts to items in the world. Instead of this objectivity, it substitutes the notions of the *trustworthiness* or *reliability* of people's estimates. Secondly, the theory does not allow for the possibility that something which expresses one's attitude or indicates an illocutionary force can also at the same time assert something about this; that is, that expressing an attitude and stating a fact are not incompatible accomplishments of the same utterance. Someone who says that so-and-so is approximately 500 yards long or that it is conceivably a danger is no doubt expressing a more guarded opinion about the actual length or the existence of a danger than someone who says it is 500 yards long or is a danger, but he is also stating a fact, namely that the length of the thing approximates to 500 yards and that it is conceivable that it is a danger. In saying that something is good we are expressing our approval of it or commending it by pointing out that it meets all the requirements made of it, and in saying that something is probable we are expressing our guarded acceptance of it by pointing out that the factors in favour of it outweigh those against it. Thirdly, to point out that saying 'It is probable that p' expresses a guarded assertion that p no more tells us what it is for it to be probable that p than to point out that saying 'I know that p' gives others my word or my authority for saying that p tells us what it is to know that p or than to point out that saying 'That is good' commends something tells us what it is for something to be good. The question *what one does* when one says something is not the question *what one says* when one says something. Finally, the theory does not see that there can be a third contrast beside the traditional contrast of subjective with objective and its own proposed contrast of illocutionary with assertive, namely, the contrast of absolute with relative.

We saw in detail in the earlier chapters that, if anything is modally characterizable, it is so because in a certain situation describable in terms of alternatives relative to a given end (e.g. for X, in order to V, because of Z, if W, as an A) and viewed under a certain aspect (e.g. physically, logically, legally, morally), it is an alternative which is open (can, may, possible), favoured (probable), required (need), owing (ought), or is the only one (necessary, must, obliged). In short, modal concepts do not signify particular items either in the world or in our minds, but the relation of one item to others in a situation.[31]

This relative nature of the modals also explains why they are 'referentially opaque'. By this is meant that, for example, though the most rabid misogynist in the Faculty is necessarily against 'Women's Lib', it does not follow that the Professor of Classics is necessarily against 'Women's Lib', even though the Professor of Classics is in fact the most rabid misogynist in the Faculty. Still less does it follow that because 'The most rabid misogynist in the Faculty is against Women's Lib' expresses what philosophers call a necessary truth, therefore, 'The Professor of Classics is against Women's Lib' expresses a necessary truth.[32] This is not, of course, to deny—what is true but irrelevant—that *if* the most rabid misogynist in the Faculty is against Women's Lib and *if* the Professor of Classics is the most rabid misogynist in the Faculty, then the Professor of Classics is necessarily against 'Women's Lib'—though even here 'The Professor of Classics is against Women's Lib' does not express a necessary truth. If X is necessarily (possibly, obligatorily) Y, it is so relative to something which could be expressed as 'If W is Z' or 'in virtue of being X', etc. It is because my colleague is a misogynist and not because he is a professor of classics that he is necessarily opposed to 'Women's Lib' just as it is because the planets number nine and not merely because they have some number that the number of the planets is necessarily greater than seven and that it is necessarily true to say that the number of the planets is greater than seven—though 'The number of the planets is greater than seven', unlike 'Nine is greater than seven', does not express what philosophers call a necessary truth. Similarly it may be because someone is a philosopher, and not because he is over 50, that he is characteristically, unusually, presumably or probably, absent-minded; while it is because he is over 50 and not because he is a philosopher that he is characteristically, unusually, presumably or probably, conservative. Hence, nothing is necessarily or possibly, any more than it is characteristically, unusually, presumably or probably, so-and-so in itself, but only in virtue of being such-and-such. To use an old-fashioned expression, one could say that something is in one of these ways so-and-so *qua* being such-and-such. The point here is exactly the same as that illustrated by the fact that though doing A may be doing B, e.g. setting fire to a bundle may be setting fire to one's passport, intentionally or knowingly doing the former does not imply intentionally or knowingly doing the latter. *Qua* setting fire to a bundle, what I did was intentional, and *qua* setting fire to my passport, it was not.

Recognition of this general point is sometimes the source of another modern variant on subjectivism, which arises in the following way. First, the point is said, rather misleadingly, to show that the attribution of such characteristics as those expressed by the words 'possibly', 'necessarily', 'presumably', 'probably', 'intentionally', etc., 'depends on the manner of referring to the object'.[33] I say 'rather misleadingly' since it is in so far as the object *is* a such and such—and not merely because it is *referred* to as a such-and-such—that it is modally qualifiable. From this misleading assertion it is then wrongly concluded that, e.g., 'being necessarily or possibly thus and so is in general not a trait of the object concerned'[34] and that 'necessity is in talk, not in things'.[35] A similar conclusion is that, e.g., 'probability does not attach to things, only to propositions and propositional functions'[36] or that setting fire to one's passport differs from setting fire to a bundle not in any feature of the event itself but only in 'a feature of the description of an event'.[37] All these remarks, which are, perhaps, variations on Wittgenstein's thesis that ' "Essential" is never the property of an object, but the mark of a concept',[38] seem to me to distort the truth that modal characteristics, though relative, are objective.

The relative nature of the modals also explains why, as we saw earlier, 'possible' does not ordinarily go with the vast majority of nouns or noun phrases, such as 'man', 'flat tyre', 'owl in a tree', since these do not ordinarily signify something which something else can or may in certain circumstances serve as or amount to. It explains, further, why conflicting obligations are so common since one of them arises in relation to one set of circumstances and the other in relation to another.

The traditional objective view, while appreciating that modality is independent of us, has confused the relational setting either with one of the items in the setting or with a mysterious extra item. It has assumed, for instance, that probability is the frequency of a certain kind of occurrence or its propensity to occur, that necessity is a power in that which necessitates, that need is a state of deficiency, that possibility is an ability, that what 'ought' indicates is either one of the properties, like maximization of happiness, in virtue of which certain things ought to be, or a peculiar non-natural property. Subjectivism, whether in the traditional form of positing a mental property or object or in the modern forms either of the distinction between the expressive and the assertive or of the distinction between language and the world, has concluded from the mistakes of traditional objectivism that there is nothing in the world indicated by modal concepts.[39] In a similar way, the mistake of traditional objectivism in ethics of supposing that value-concepts indicate either natural or non-natural items in the world led either to traditional subjectivist theories or to modern illocutionary theories, (e.g. emotivism, prescriptivism). But *good*, *right*, etc., are contextual in a similar way to modal concepts. To be a 'good' X or 'the right' X does not mean to be an X which possesses certain natural properties—namely, those which make it a good X or the right X—nor to be an X which possesses some extra non-natural property, nor to be an X which arouses approval in us, nor an X which we prescribe. It means to be an X which has whatever are the natural properties designated as desirable in Xs. The quality of meeting in various ways certain requirements, which is what being good, right, suitable, eligible, or whatever, is, is different from, but no less objective than, those qualities which enable anything to meet these requirements. Philosophers were, I think, right to suggest that modal concepts are analogous to value concepts—though not right to assimilate[40] the former to the latter—and right to think that the analyses of both kinds of concepts, whether in the traditional categories of objective and subjective or in the modern categories of expressive and assertive, stand and fall together. They were right to suggest that in both there is a contrast of what has been called their meaning, or force, with their criteria. Because of this contrast, there is an answer to the question what 'makes' something possible or necessary, right or good, eligible or suitable, but not to the question what makes it red or round, old or oriental. As the presence of the naturalistic fallacy warns us, it is always possible to ask sensibly whether something possessed of such-and-such particular features is necessary, possible, probable or what ought to be, or is what is good or right, whatever these features are. It is possible to know what is meant by saying that so-and-so is necessary, possible, obligatory, needed or ought to be, just as it is possible to know what is meant by saying that it is good, without having any idea of what characteristics it actually has, because one does not know what are the criteria for necessity, or whatever, in so-and-so. Contrariwise, it is possible to know what features so-and-so has without knowing that it is, therefore, necessary or possible, good or valuable. Though two things could be alike in every other respect but differ in colour or shape or age, two things could not be identical in every other respect but differ in a modal or a value characteristic.

The usual suggestions for an analysis of modal and value concepts that would account for these characteristics seem to me to be mistaken for the reasons I have given. I hope to have provided for each modal concept an alternative analysis, in terms of the relation between the modalized item and the other items in the situation, which allows modal notions—and the same would apply to evaluative notions—an objectivity free from traditional objections. I hope also to have shown that my detailed examination of the particular modals supports this analysis.

NOTES

1. (1951b), pp. 8, 25.
2. Cp. Prior (1952); Kneale (1962b).
3. von Wright admits this in (1951b), p. 25.
4. *Pace* Plantinga (1969), pp. 235, 238, part of whose reason for this view is a failure to see the ordinary use of 'It is necessarily true that X is F', in which it is not synonymous with 'X is F is a necessary truth', but expresses something logically equivalent to 'X is necessarily F'. Because he wrongly allows the former synonymity, he wrongly denies the latter logical equivalence.

5. E.g. (1951b); contrast Hughes and Cresswell, pp. 183–8.
6. As the medievals may have tried to do, e.g. Prior (1952), p. 175; (1962), part III, ch. 1; Kneale (1962b), pp. 624–6.
7. Followed by R. Campbell and, perhaps, by Kneale (1962b), pp. 626–7, and Plantinga (1974), who sometimes seem to suggest that to be predicable of a whole is necessarily to be predicable *de dicto*. Indeed many modern logicians seem to use the *de dicto/de re* distinction as a distinction in the scope of a quantifier.
8. Cp. Kneale (1962a), pp. 117–18, 236–7.
9. E.g. Toulmin, Harré, Rescher, Plantinga (1974) and nearly all modal logicians.
10. E.g. ch. 4.
11. Ch. 5.
12. Cp. White (1975).
13. p. 321.
14. p. 31.
15. Entirely in (1951a) and partly in (1951b).
16. (1951b).
17. It is not quite clear to me whether von Wright ever distinguished between deontic qualifications of actions and qualifications of descriptions, or names, of actions or only between names of actions and descriptions of states of affairs, cp. (1968), pp. 13–15, 91. At one point, p. 16, he says 'Instead of proposition we can also say "possible state of affairs".'
18. Cp. Kneale (1962a), p. 548.
19. D. G. Brown, § 3.9.
20. Edgley, p. 134 and § 4.10.
21. Edgley, § 4.10.
22. D. G. Brown, § 3.3, 76.
23. Edgley, § 4.10; cp. D. G. Brown, § 3.3. In the final paragraph of this section, on the contrary, Brown seems to argue that neither he nor Hume is denying that necessity applies to events, but only giving an analysis of such necessity.
24. Edgley, § 3.22.
25. *Treatise* I.iii. 14.
26. *Treatise* III.i. 1–2.
27. Lyons, p. 307.
28. Toulmin, chs. 1 and 2; cp. Harré.
29. Toulmin, pp. 32–3, 69.
30. N. Kemp Smith, *The Philosophy of David Hume* (1941).
31. Fogelin's analysis of alethic modalities, ch. 2, and of value concepts, ch. 7, in terms of 'warrants' sees this, but he too assumes that modals, like warrants, qualify propositions.
32. Quine (1953), ch. VIII assumes that 'X is necessarily Y' is equivalent to ' "X is Y" is a necessary truth', whereas it is equivalent only to the quite different 'It is necessarily true that X is Y'.
33. Quine (1953), p. 148.

34. *Ibid.*
35. Quine (1966), p. 174; cp. Kneale (1962b), p. 629.
36. Lucas (1970), p. 187.
37. D. Davidson, 'Agency' in *Agent, Action and Reason*, ed. by Binkley, Bronaugh and Marras (1971), p. 22.
38. *Remarks on the Foundations of Mathematics* (1956), I § 73.
39. E.g. Toulmin, pp. 62–71.
40. E.g. Toulmin, Brown, Edgley and Peters (1958).

Bibliography

Aaron, R. I., 'Feeling Sure', *Proc. Arist. Soc.*, Suppl. 30 (1956), 1–13.

Anscombe, G. E. M., 'Aristotle and the Sea-Battle', *Mind*, 65 (1956), 1–15.

—— *Intention* (Oxford, 1957).

—— 'On Brute Facts', *Analysis*, 18 (1958), 69–72.

—— 'Modern Moral Philosophy', *Philosophy*, 33 (1958).

Archambault, R. D., 'The Concept of Need and its relation to certain aspects of Educational Theory', *Harvard Educational Review* (1957), 38–62.

Aune, B., 'Abilities, Modalities and Free-Will', *Phil. & Phenom. Research* 23 (1962/3), 397–413.

—— 'Hypotheticals and "Can": Another Look', *Analysis*, 27 (1967), 191–5.

—— 'Free will, "can" and ethics; a reply to Lehrer', *Analysis*, 30 (1970), 77–83.

Austin, J. L., *Philosophical Papers* (Oxford, 1961).

—— *Sense and Sensibilia* (Oxford, 1962).

Ayer, A. J., 'Freedom and Necessity' (1946), reprinted in *Philosophical Essays* (London, 1954), 271–84.

—— *The Problem of Knowledge* (Harmondsworth, 1956).

—— 'Two Notes on Probability', reprinted in *The Concept of a Person* (London, 1963), 188–208.

—— *Probability and Evidence* (London, 1972).

Ayers, M. R., 'Austin on "could" and "could have"', *Phil Quart.*, 16 (1966), 113–20.

—— *The Refutation of Determinism* (London, 1968).

Baier, K., *The Moral Point of View* (Ithaca, New York, 1958).

—— 'Could and Would', *Analysis*, 23 Supplement (1963), 20–9.

—— 'Moral Obligation', *American Phil. Quart.*, 6 (1966), 210–26.

Bales, R. E., 'Utilitarianism; Overall Obligatoriness and Deontic Logic', *Analysis*, 32 (1972), 203–5.

Barry, R. M. and Rees, W. J., 'Public Interest', *Proc. Arist. Soc.*, Suppl. 38 (1964), 1–38.

Baumeister, A., Hawkins, W. F. and Cromwell, R. L., 'Need States and Activity Levels', *Psych. Bull.*, 61 (1964), 438–53.

Bergmann, G., 'The Philosophical Significance of Modal Logic', *Mind*, 69 (1960), 466–85.
Benn, S. I., '"Interests" in Politics', *Proc. Arist. Soc.*, 60 (1960), 123–140.
Benn, S. I. and Peters, R. S., *Social Principles and the Democratic State* (London, 1959).
Bernoulli, J., *Ars Conjectandi* (1713).
Black, M., 'Possibility', *J. of Phil.*, 57 (1960), 117–26.
—— 'The Gap between "Is" and "Should"', *Phil. Review*, 73 (1964), 165–81.
—— 'Probability', in *Encyclopaedia of Philosophy*, edited by P. Edwards VI (New York, 1967), 464–79.
Blom, S., 'Concerning a Controversy on the meaning of "Probability"', *Theoria*, 21 (1955), 65–98.
Boyd, J. and Thorne, J. P., 'The deep grammar of modal verbs', *J. of Linguistics*, 5 (1969), 57–74.
Braine, D., 'Varieties of Necessity', *Proc. Arist. Soc.*, Suppl. 46 (1972), 139–70.
Braithwaite, R. B., 'On Unknown Probabilities', in *Observation and Interpretation*, edited by S. Korner (London, 1957).
Braybrooke, D., 'Let Needs Diminish that Preferences May Prosper', *Studies in Moral Philosophy, American Phil. Quart.*, Monograph Series I (1968), 86–107.
Broad, C. D., *Five Types of Ethical Theory* (London, 1930).
—— 'Determinism, Indeterminism and Libertarianism', *Ethics and the History of Philosophy* (London, 1952), 195–217.
Brown, D. G., *Action* (London, 1968).
Brown, J. S., *The Motivation of Behaviour* (New York, 1961).
Byrne, E. F., *Probability and Opinion* (The Hague, 1968).
Campbell, C. A., 'Is "Free will" a Pseudo-Problem?', *Mind*, 60 (1951), 441–65.
Campbell, R., 'Modality *De Dicto* and *De Re*', *Australas., J. of Phil.*, 42 (1964), 345–359.
Carnap, R., 'The Two Concepts of Probability', *Phil. & Phenom. Research* (1944/5).
—— *Meaning and Necessity* (Chicago, 1947, and 2nd 1956).
—— *Logical Foundations of Probability* (Chicago, 1962).
Castañeda, H. N., 'The Logic of Change, Action and Norms', *J. of Phil.*, 62 (1961), 337–44.
—— 'A problem for Utilitarianism', *Analysis*, 28 (1968), 141–2.
Chisholm, R. M., *Perceiving* (Ithaca, New York, 1957).
—— 'J. L. Austin's Philosophical Papers', *Mind*, 73 (1964), 20–5.
—— 'He could have done otherwise', *J. of Phil.*, 64 (1967), 409–17.
Day, J. P., *Inductive Probability* (London, 1961).
Dearden, R. F., '"Needs" in Education', *Br. J. of Educ. Studies*, 14 (1966), 577.

—— *The Philosophy of Primary Education* (London, 1968).
Diggs, A. J., 'A Technical Ought', *Mind*, 69 (1960), 301–17.
Dore, C., 'On the meaning of could have', *Analysis*, 23 (1962), 41–3.
—— 'On being able to do otherwise', *Phil. Quart.*, 16 (1966), 137–45.
Edgley, R., *Reason in Theory and Practice* (London, 1969).
Ehrman, M. E., *The Meaning of the Modals in present day American English* (The Hague, 1966).
Ewing, A. C., *The Definition of Good* (London, 1947).
—— 'May can-statements be analysed deterministically?', *Proc. Arist. Soc.*, 64 (1964), 157–76.
Findlay, J. N., 'Probability without Nonsense', *Phil. Quart.*, 2 (1952), 218–39.
de Finetti, B., 'Foresight: Its Logical Laws, Its Subjective Sources' (1937), reprinted in H. E. Kyburg and H. C. Smokler, *Studies in Subjective Probability* (New York, 1964).
Firth, R., 'The Anatomy of Certainty', *Phil. Rev.*, 76 (1967), 3–27.
Fogelin, R. J., *Evidence and Meaning* (London, 1967).
Foot, P. R., 'Moral Beliefs', *Proc. Arist. Soc.*, 59 (1959), 83–104.
Frankena, W., 'Obligation and Ability', in *Philosophical Analysis*, edited by M. Black (Ithaca, New York, 1950), 157–75.
Frankfurt, H. G. 'Philosophical Certainty', *Phil. Rev.*, 71 (1962) 303–27.
Gallop, D., 'On being determined', *Mind*, 71 (1962), 181–96.
—— 'Ayers on "Could" and "Could have" ', *Phil. Quart.*, 17 (1967), 216–24.
Gauthier, D. P., *Practical Reasoning* (Oxford, 1963).
Gert, B., and Martin, J. A., ' "What a man does he can do"?,' *Analysis*, 33 (1973), 168–73.
Gibbs, B., 'Real Possibility', *American Phil. Quart.*, 7 (1970), 340–8.
Gilbert, M., 'The Abilities of Prescriptivism', *Analysis*, 32 (1972), 141–4.
Greenbaum, S., *Studies in English Adverbial Usage* (London, 1969).
Grice, G. R., *The Grounds of Moral Judgment* (Cambridge, 1967).
Hacking, I., *The Logic of Statistical Inference* (London, 1965).
—— 'Possibility', *Phil. Rev.*, 76 (1967), 143–8.
Hamlyn, D. W., 'On Necessary Truth', *Mind*, 70 (1961), 514–25.
Hampshire, S. N., 'Freedom of the Will', *Proc. Arist. Soc.*, Suppl. 25 (1951), 161–78.
—— *Freedom of the Individual* (London, 1965).
Hare, R. M., *The Language of Morals* (Oxford, 1952).
—— *Freedom and Reason* (Oxford, 1963).
Harré, R., 'Modal Expressions in Ordinary and Technical Language', *Australas. J. of Phil.*, 37 (1959), 41–56.
Hart, H. L. A., 'Legal and Moral Obligation', *Essays in Moral Philosophy*, edited by A. I. Melden (Seattle, Washington, 1958).
—— *The Concept of Law* (Oxford, 1961).
Henderson, G. P., ' "Ought" implies "Can" ', *Philosophy*, 41 (1966), 101–12.

Heidelberger, H., 'Knowledge, Certainty and Probability', *Inquiry*, 6 (1963), 242–50.
Hintikka, J., *Knowledge and Belief* (New York, 1962).
Hinton, J. M., 'Hoping and Wishing', *Proc. Arist. Soc.*, Suppl. 44 (1970), 71–88.
Honoré, A. M., 'Can and Can't', *Mind*, 73 (1964), 463–79.
Hughes, G. E. and Cresswell, M. J., *An Introduction to Modal Logic* (London, 1968).
Hull, C. L., *Principles of Behaviour* (New York, 1943).
Hunter, J. F. M., 'Aune and others on Ifs and Cans', *Analysis*, 28 (1968), 107–9.
Joos, M., *The English Verb* (Madison, 1964).
Kaufman, A. S., 'Ability', *J. of Philosophy*, 60 (1963), 547–50.
—— 'Moral responsibility and the use of "could have"', *Phil. Quart.* 12 (1962), 120–8.
Kearney, R. J., 'Meaning and Implication: Other Thoughts', *Analysis*, 33 (1972), 47–50.
Kenny, A., *Action, Emotion and Will* (London, 1963).
Keynes, J. M., *A Treatise on Probability* (London, 1921).
King-Farlow, J., 'Toulmin's analysis of probability', *Theoria*, 29 (1963), 12–26.
Kneale, W. C., *Probability and Induction* (Oxford, 1949).
—— *The Development of Logic* (Oxford, 1962).
—— 'Modality *De Dicto* and *De Re*', in *Logic, Methodology and Philosophy of Science*, edited by Nagel, Suppes and Tarski (Stanford, 1962), 622–33.
Komisar, B. P., '"Need" and the Needs-Curriculum' in *Language and Concepts in Education*, edited by B. O. Smith and R. H. Ennis (1961).
Kordig, C. R., 'Moral Weakness and Self-Reference', *Analysis*, 32 (1971), 11–12.
Kyburg, H. E., *Probability and the Logic of Rational Belief* (Connecticut, 1961).
Kyburg, H. E. and Smokler, H. C. (ed.) *Studies in Subjective Probability* (New York, 1964).
Lebrun, Y., *'Can' and 'May' in present-day English* (Brussels, 1965).
Lehrer, K., 'Ifs, cans and causes', *Analysis*, 20 (1960), 122–4.
—— 'Cans and conditionals: A Rejoinder', *Analysis*, 22 (1961), 23–4.
—— 'An empirical disproof of determinism', in *Freedom and Determinism* (New York, 1966).
—— 'Cans without Ifs', *Analysis*, 29 (1968), 29–32.
Levi, I., *Gambling with Truth* (London, 1967).
Lewis, C. I., *Analysis of Knowledge and Valuation* (La Salle, Illinois, 1946).
Locke, D., 'Ifs and Cans revisited', *Philosophy*, 37 (1962), 245–56.
Lucas, J. R., *The Concept of Probability* (Oxford, 1970).

Lyons, J., *Introduction to Theoretical Linguistics* (Cambridge, 1968).
McGuinness, B. F., ' "I know what I want" ', *Proc. Arist. Soc.*, 57 (1957), 305–20.
Mackie, J. L., *Truth, Probability and Paradox* (Oxford, 1973).
Malcolm, N., 'Certainty and Empirical Statements', *Mind*, 51 (1942), 18–46.
—— *Knowledge and Certainty* (New Jersey, 1963).
Margolis, J., 'The Analysis of "Ought" ', *Australas. J. of Phil.*, 48 (1970), 44–53.
—— *Values and Conduct* (Oxford, 1971).
Maslow, A. H., *Motivation and Personality* (New York, 1954).
Matthews, G. R. and Cohen, S. M., 'Wants and Lacks', *J. of Phil.*, 64 (1967), 455–6.
Mavrodes, S. I., ' "Is" and "ought" ', *Analysis*, 25 (1964), 42–4.
Mayo, B., 'On the Lehrer-Taylor Analysis of "Can" statements', *Mind*, 77 (1968), 271–8.
Melden, A. I., *Free Action* (London, 1961).
Mellor, D. H., 'Chance', *Proc. Arist Soc.*, Suppl. 43 (1969), 11–36.
—— *The Matter of Chance* (Cambridge, 1971).
Miller, J. D. B., *The Nature of Politics* (Harmondsworth, 1962).
Montefiore, A. C., ' "Ought" and "Can" ', *Phil, Quart.* 8 (1958), 24–40.
Moore, G. E., *Principia Ethica* (Cambridge, 1903).
—— *Philosophical Studies* (London, 1922).
—— *Some Main Problems of Philosophy* (London, 1953).
—— *Philosophical Papers* (London, 1959).
—— *Commonplace Book* (London, 1962).
de Morgan, A., *Formal Logic* (London, 1847).
Nagel, E., *Principles of the Theory of Probability*, in *International Encyclopaedia of Unified Science*, Vol. 1, No. 6 (Chicago, 1939).
Nielsen, K., 'Morality and Needs', in *The Business of Reason*, edited by J. J. MacIntosh and S. Coval (London, 1969), 186–206.
Nolan, Rita 'The Parsing of "Possible" ', *J. of Phil.*, 69 (1972), 157–68.
Nowell-Smith, P. H., *Ethics* (Harmondsworth, 1954).
—— 'Ifs and Cans', *Theoria*, 26 (1960), 85–101.
O'Connor, D. J., *Free Will* (London, 1971).
—— 'Possibility and Chance', *Proc. Arist. Soc.*, Suppl. 34 (1960), 1–24.
Osborn, J. M., 'Austin's Non-Conditional Ifs', *J. of Phil.*, 62 (1965), 711–5.
Page, E., 'Senses of "Obliged" ', *Analysis*, 33 (1972), 42–6.
—— 'On being obliged', *Mind*, 82 (1973), 283–8.
Palmer, F. R., *A Linguistic Study of the English Verb* (London, 1965).
Pap, A., and Scriven, M., 'Are dispositional statements hypothetical?', in *Minnesota Studies in the Philosophy of Science* II (Minneapolis, 1958).
Pears, D. F., 'Ifs and Cans', reprinted in *Essays on J. L. Austin*, edited by Berlin *et al.* (Oxford, 1973), 90–140.
Peirce, C. S., 'Notes on the Doctrine of Chances' (1910), in *Collected*

Papers of C. S. Peirce, edited by Hartshorne and Weiss II (1932), 404–14.
Peters, R. S., *The Concept of Motivation* (London, 1958).
—— *Ethics and Education* (London, 1966).
Peters, R. S. and Benn, S. H. *Social Principles and the Democratic State* (London, 1959).
Peters, R. S. and Hirst, P. H., *The Logic of Education* (London, 1970).
Plantinga, A., '*De Re et De Dicto*', *Nous*, 3 (1969), 235–58.
—— *The Nature of Necessity* (Oxford, 1974).
Popper, K. R., 'The Propensity Interpretation of the Calculus of Probability and the Quantum Theory', in *Observation and Interpretation*, edited by S. Korner (London, 1957), 65–70.
—— 'The Propensity Interpretation of Probability', *Brit. J. Phil. Sci.*, 10 (1959), 25–42.
—— 'Quantum Mechanics without "The Observer"', in *Quantum Theory and Reality*, edited by M. Bunge (Berlin, 1967).
Prichard, H. A., *Duty and Ignorance of Fact* (Oxford, 1932).
—— *Moral Obligation* (Oxford, 1949).
Prior, A. N., 'Modality *de dicto* and modality *de re*', *Theoria*, 18 (1952), 174–80.
—— *Formal Logic* (Oxford, 2nd, 1962).
Quine, W. V., 'Reference and Modality', in *From a Logical Point of View* (Cambridge, Massachusetts, 1953).
—— *Word and Object* (Cambridge, Massachusetts, 1960).
—— 'Necessary Truth', in *Ways of Paradox* (New York, 1966), 48–56.
Raab, F. V., 'Freewill and the ambiguity of "could"', *Phil. Rev.*, 64 (1955), 60–72.
Radford, C., 'Hoping and Wishing', *Proc. Arist. Soc.*, Suppl. 44 (1970), 51–70.
Ramsey, F. P., *The Foundations of Mathematics* (London, 1931).
Rescher, N., *Topics in Philosophical Logic* (Dordrecht, 1968).
Robinson, R., 'Ought and Ought Not', *Philosophy*, 46 (1971), 193–202.
—— 'The Concept of Knowledge', *Mind*, 80 (1971), 17–28.
Rollins, C. D., 'Certainty', in *Encyclopaedia of Philosophy*, edited by P. Edwards II (New York, 1967), 67–71.
Ross, W. D., *The Right and the Good* (Oxford, 1930).
—— *Foundations of Ethics* (Oxford, 1939).
Russell, B., *Human Knowledge* (New York, 1949).
Ryle, G., *The Concept of Mind* (London, 1949).
Scarrow, D. S. 'On an analysis of "could have"', *Analysis*, 23 (1963), 118–20.
Searle, J. R., *Speech Acts* (Cambridge, 1969).
Sesonske, A., *Value and Obligation* (Berkeley, California, 1957).
Shaw, P. D., '"Ought" and "Can"', *Analysis*, 25 (1965), 196–7.
Skinner, B. F., *Science and Human Behaviour* (New York, 1953).
Sloman, A., '"Ought" and "Better"', *Mind*, 79 (1970), 385–94.

Smith, J. W., 'Impossibility and Morals', *Mind*, 70 (1961), 362–75.
Sparshott, F. E., *An Enquiry into Goodness* (Toronto, 1958).
Strang, C., 'Aristotle and the Sea-Battle', *Mind*, 69 (1960), 447–65.
Taylor, P. W., ' "Need" Statements', *Analysis*, 19 (1959), 106–11.
Taylor, R., 'The Problem of Future Contingencies', *Phil. Rev.* 66 (1957).
—— 'I can', *Phil. Rev.* 69 (1960), 78–89.
—— Introduction to *The Ontological Argument*, edited by A. Plantinga (New York, 1965).
—— *Action and Purpose* (New Jersey, 1966).
Thalberg, I., 'Abilities and Ifs', *Analysis*, 22 (1962), 121–6.
—— 'Freedom of Action and Freedom of Will', *J. of Phil.*, 61 (1964), 405–14.
—— 'Austin on Abilities', in *Symposium on J. L. Austin*, edited by K. T. Fann (London, 1969), 182–204.
Toulmin, S. E., *The Uses of Argument* (Cambridge, 1958).
Twadell, W. F., *The English Verb Auxiliaries* (Providence, Rhode Island, 1963).
Urmson, J. O., 'Two of the senses of "Probable" ', *Analysis*, 8 (1948), 9–16.
Venn, J., *The Logic of Chance* (London, 1888).
Wertheimer, R., *The Significance of Sense* (Ithaca, New York, 1972).
White, A. R., *G. E. Moore: A Critical Exposition* (Oxford, 1958).
—— *Attention* (Oxford, 1964).
—— 'Meaning and Implication', *Analysis*, 32 (1971), 26–30.
—— 'The Propensity Theory of Probability', *Brit. J. Phil. Sci.* 23 (1972), 35–43.
—— 'Belief as a propositional attitude', in *Russell*, edited by G. W. Roberts (1975).
White, M., 'On what could have happened', *Phil. Rev.*, 77 (1968), 73–89.
Whiteley, C. H., 'Can', *Analysis*, 23 (1963), 91–3.
Wilson, P. S., *Interest and Discipline in Education* (London, 1971).
Wittgenstein, L., *Philosophical Investigations* (Oxford, 1953).
—— *The Blue and Brown Books* (Oxford, 1958).
—— *On Certainty* (Oxford, 1969).
von Wright, G. H., 'Deontic Logic', *Mind*, 60 (1951), 1–15.
—— *An Essay in Modal Logic* (Amsterdam, 1951).
—— *A Treatise on Induction and Probability* (London, 1951).
—— *The Varieties of Goodness* (London, 1963).
—— *Norm and Action* (London, 1963).
—— 'An Essay in Deontic Logic and the General Theory of Action', *Acta Philosophica Fennica* 21 (Amsterdam, 1968).
Zink, S., *The Concepts of Ethics* (London, 1962).

Index

(Names occurring only in references and bibliography are not included.)